Markus Kaltenborn • Markus Krajewski •
Heike Kuhn
Editors

Sustainable Development Goals and Human Rights

Springer Open

Editors
Markus Kaltenborn
Faculty of Law
Ruhr University Bochum
Bochum, Germany

Markus Krajewski
Faculty of Law
University of Erlangen-Nuremberg
Erlangen, Germany

Heike Kuhn
Division Human Rights, Gender Equality,
Inclusion of Persons with Disabilities
Federal Ministry of Economic Cooperation
and Development
Bonn, Germany

ISSN 2509-2960 ISSN 2509-2979 (electronic)
Interdisciplinary Studies in Human Rights
ISBN 978-3-030-30468-3 ISBN 978-3-030-30469-0 (eBook)
https://doi.org/10.1007/978-3-030-30469-0

This Springer imprint is published by the registered company Springer Nature Switzerland AG.
The registered company address is: Gewerbestrasse 11, 6330 Cham, Switzerland

Preface

Claiming that human rights and the United Nations Sustainable Development Goals (SDGs) have much in common and contribute to their mutual implementation seems like stating the obvious. Yet, more often than not discussions about human rights and about the SDGs are held in different circles and by different stakeholders.

Bringing together these stakeholders was the goal of a Conference in June 2017 at the Federal Ministry for Economic Cooperation and Development in Bonn, Germany, organized by the Institute of Development Research and Development Policy (IEE) of the Ruhr-University Bochum and the Centre for Human Rights Erlangen-Nürnberg (CHREN) in collaboration with the Federal Ministry for Economic Cooperation and Development. We would like to acknowledge the financial support of the State Chancellery of North Rhine-Westphalia and of the Federal Ministry for Economic Cooperation and Development, which made this conference possible. We also would like to thank Melina Anagnostopoulou Ribeiro of Ruhr-University Bochum as well as Walter Gemein and Stefanie Mostert of the Federal Ministry for Economic Cooperation and Development for their great efforts and substantive help in organizing this conference.

This volume collects some of the papers presented at this conference but also goes beyond the discussions. We gratefully acknowledge a grant of the Federal Ministry for Economic Cooperation and Development, which allowed us to publish the present volume as an open access document. We believe this is important in light of the analyses and debates of the contributions of this volume.

We hope that this volume contributes to mutual learning and reinforcement of human rights and the SDGs.

Bochum, Germany	Markus Kaltenborn
Erlangen, Germany	Markus Krajewski
Bonn, Germany	Heike Kuhn
July 2019	

Contents

About the Editors and Contributors

Alan Boyle is Emeritus Professor of Public International Law at the University of Edinburgh. He is a specialist in public international law, especially international environmental law, law of the sea, international law-making and international litigation. He has worked regularly as a consultant or as counsel for various governments, international organizations and NGOs on general international law, international environmental law and law of the sea. Inter alia he has appeared as counsel in six cases brought under the 1982 UN Convention on the Law of the Sea. He is one of the leading practitioners in that field and in international environmental litigation, where he has been counsel in many of the leading cases. Educated at Oxford University, he has also taught at the University of London (Queen Mary College); University of Texas Law School; William and Mary College Law School, Virginia; the University of Paris and LUISS in Rome. He is a barrister and practices occasionally in the International Court of Justice and other international tribunals, mainly in environmental and law of the sea cases.

Delanyo Dovlo was the Director of Health Systems and Services Cluster in the WHO Regional Office for Africa (AFRO). Previously he served as WHO Country Representative in Rwanda and Health Systems Adviser at WHO headquarters. A Ghanaian public health physician with over 30 years of clinical and public health experience, he has been a consultant to various countries on Health Systems and Health Sector Reforms. He has an MB CHB from the University of Ghana, an MPH from University of Leeds, a Membership of the West Africa College of Physicians and is a Fellow of the Ghana College of Physicians & Surgeons.

Markus Kaltenborn is Professor of Public Law at the Ruhr-University Bochum, Faculty of Law. He is Director of the Institute of Development Research and Development Policy (IEE) at Ruhr University Bochum, member of the German Institute for Human Rights (DIMR e.V.) and of the supervisory board of Oxfam Germany e.V; in recent years he has also been Visiting Lecturer at the Institute for Social Development of the University of the Western Cape (Cape Town). His main areas of research are Health Law, Social Protection Law, the Law of Development Cooperation and Human Rights Law. He is (together with Katja Bender and Christian Pfleiderer) editor of the book Social Protection in Developing Countries. Reforming Systems (Routledge 2013) as well as (together with Philipp Dann und Stefan Kadelbach) of the edited volume Entwicklung und Recht (Development and Law, Nomos 2014). Furthermore he is the author of a study on Social Rights and International Development (Springer 2015).

Markus Krajewski holds the Chair in Public Law and Public International Law at the University of Erlangen-Nürnberg. His research focusses on international economic law, human rights, European external relations and the law of public services. He heads the Interdisciplinary Research Centre for Human Rights Erlangen-Nürnberg (CHREN) and is one of the programme directors of the Master programme in Human Rights at the University of Erlangen-Nürnberg. He also chairs the Board of Trustees of the German Institute for Human Rights and is Secretary-General of the German Branch of the International Law Association.

Heike Kuhn is Head of Division "Human Rights, gender equality, inclusion of persons with disabilities" at the Federal Ministry for Economic Cooperation and Development (BMZ), Bonn/Berlin). She has a background of more than 25 years in European and multilateral development cooperation, human rights, financing and programming. Ms. Kuhn has long-standing experience in leading and managing multilateral negotiations at European and UN-level, lately as Executive Board Director to the International Fund of Agricultural Development and Alternate Representative at the Permanent Representation of the Federal Republic of Germany to the International Organizations in Rome/Italy. Ms. Kuhn holds a Ph.D. in Administrative Sciences from the University of Speyer (Germany) on "The social dimension of the European Community". She is a fully qualified lawyer, having studied in Marburg and Freiburg/Germany, Moscow/USSR, Los Angeles/USA and Genoa/Italy.

Liora Lazarus BA (UCT), LLB (LSE), DPhil (Oxon), is an Associate Professor in Law, and a Fellow of St. Anne's College, University of Oxford. Her primary research interests are in comparative human rights, security and human rights, comparative theory and comparative criminal justice. Born and raised in South Africa, she studied African Economic History at the University of Cape Town and Law at the London School of Economics and Political Science. From 1994 to 1995 she was a Fellow of the Max Planck Institute for Foreign and International Criminal Law, Freiburg, Germany. She came to Oxford in 1995 to write her doctorate at Balliol College, after which she went on to become a law fellow at St Anne's College. Dr. Lazarus's various academic publications are in the area of prisoners' rights, criminal justice and security and human rights. She has completed public reports for the UK Ministry of Justice on balancing public protection with human rights, and on the relationship between rights and responsibilities. She has acted as an advisor to the UK Stern Review on Rape Complaints, was asked to give evidence to the Parliamentary Modern Slavery Bill Committee, and was part of an International Committee of Experts to the Basque Working Group on Treatment of Political Detainees. Dr. Lazarus co-founded and is actively involved in the work of Oxford Pro Bono Pulico. She is also an Associate of the Oxford Human Rights Hub. Dr. Lazarus is an Academic Affiliate of the Bonavero Institute of Human Rights at Oxford and a Member of Oxford's Centre for Criminological Research. She sits on the Advisory Board of the Max-Planck Encyclopedia of Comparative Constitutional Law. Dr. Lazarus founded and now acts as the series editor of the Hart Studies in Security and Justice.

Jens Martens is Executive Director of Global Policy Forum (New York/Bonn) and has been the director of Global Policy Forum Europe since its foundation in 2004. Since 2011 he has coordinated the international Civil Society Reflection Group on the 2030 Agenda for Sustainable Development. From 2003 to 2009 he was member (2006–2009 Co-Chair) of the Coordinating Committee of Social Watch, a global network of several hundred NGOs working on poverty eradication and social justice. He is also a member of the Advisory Board of the Development and Peace Foundation. Previously, he was a longstanding member of the Executive Board of the German NGO World Economy, Ecology and Development (WEED). Prior to joining the staff at WEED he worked as freelance author and advisor for several NGOs and Foundations, among others the German NGO Forum on Environment and Development, the Friedrich Ebert Foundation, and the Development and Peace Foundation. From 1991 to 1992 he worked as librarian and research fellow at the German Commission for UNESCO in Bonn. He has published many articles in journals, newspapers and handbooks, has written several books and studies on development policy and UN reform, and has been co-editor of three books on German UN policy, privatization and corporate accountability. Between 2003 and 2015 he was author of the German Reality of Aid Report published annually by Deutsche Welthungerhilfe and terre des hommes Germany. He studied economics and political sciences at the University Erlangen-Nürnberg and at the Free University of Berlin.

Judy Mulusa is a Professor of Economics at Prince George's Community College (USA) and an adjunct professor at Bowie State University (USA). She also serves as a Research Associate for Claret Consulting a consulting firm involved in public private partnerships and infrastructure projects. Although her research interests are broad she has a particular interest in economic growth and development, economics or urban redevelopment and public finance. She has most recently published work on gentrification and the housing market in the United States. Previously Judy held faculty positions in her native country, Kenya at Moi University and at Masinde Muliro University of Science and Technology. She also served as an Economist with the Ministry of Finance. Judy received a B. A. in economics and political science minor from the University of Nairobi in Kenya, an M. A. in Economics from the University of Botswana and a Ph. D. in Economics from Howard University (USA).

Josephine A. Odera is the Director of ACTIL (Africa Centre for Transformative & Inclusive Leadership). Before her appointment at ACTIL, she was the Regional Director, West and Central Africa, UN Women and then briefly served as Regional Advisor, Leadership & Governance at the United Nations Women Regional Office for East and Southern Africa. Mrs. Odera holds a Doctor of Philosophy (PhD) degree in Conflict Management from the University of Ghent, Belgium having previously obtained a master's degree in Economics, a Bachelor of philosophy degree in Economics and a Bachelor of Arts degree.

She has served in the Ministry of Foreign Affairs and International Cooperation of the Republic of Kenya and also taught at the University of Nairobi, Kenya, worked in Civil Society as a Director of Programmes and most recently as Regional Director in the United Nations. As a government official, Dr. Odera represented the Government of Kenya in various meetings including the UN General Assembly, ECOSOC, African Union, regional trade and cooperation meetings such as COMESA. Dr. Odera was also a pioneer for the review of Foreign Service regulations as they affected women. She has also served as joint secretary to various bilateral commissions of the Republic of Kenya. She has served as a Senior Program consultant for UNDP in Addis Ababa and has undertaken numerous other consultancies in the Horn of Africa and the Great Lakes region.

Beate Rudolf is Director of the German Institute for Human Rights (GIHR), the National Human Rights Institution of Germany (A-status accredited under the United Nations Paris Principles). From 2016 to 2019, she also chaired the Global Alliance of National Human Rights Institutions (GANHRI). Prior to her appointment to the GIHR, she was a junior professor for public law and equality law at the faculty of law of Freie Universität Berlin and director of the research project "Public International Law Standards for Governance in Weak and Failing States" within the Research Center "Governance in Areas of Limited Statehood". Her research focuses on human rights and legal principles on state structures under public international law, European law and German constitutional law. She was Vice-President of the European Women Lawyers Association (EWLA) and is a member of the ILA Committee on Feminism in International Law as well as an advisory member of the Committee on Public International and European Law of the German Women Lawyers' Association (djb). She is a co-editor of the international CEDAW Commentary published by Oxford University Press and was awarded the Margherita-von-Brentano-Award of Freie Universität Berlin for her outstanding academic and socio-political contribution to human rights, in particular women's human rights.

Hans-Otto Sano is Director of Research at the Danish Institute for Human Rights and currently involved in three research projects at the Danish Institute for Human Rights: one on human rights-based approaches and poverty reduction, one on human rights and economics, and one on human rights research methods including how institutional theory can be applied in human rights research. Mr. Sano has worked earlier on subjects relating to governance and social accountability, on human rights indicators, and on the meaning and value added of human rights-based approaches. He has research experience from several countries in sub-Saharan Africa and from South Asia.

Christoph Scherrer is Professor of Globalization and Politics and the Executive Director of the International Center for Development and Decent Work (ICDD) at the University of Kassel, and a member of the Steering Committee of the Global Labour University. He holds PhDs in political science from the FU Berlin (Dr. habil.) and the University of Frankfurt (Dr. phil.) and a M.A. in economics (U of Frankfurt). He directs two English-language Master programs (MA Global Political Economy (GPE) and Labour Policies and Globalization (LPG)) and one PhD program. He has received the Excellency in Teaching Award of the State of Hessia and the Excellence in Development Cooperation Award from the DAAD.

Imme Scholz is a sociologist and Acting Director of the German Development Institute (Deutsches Institut für Entwicklungspolitik DIE) since 2009. Before, she headed the department on environmental policy and natural resource management at DIE, and introduced work related to climate change into the institute's research and policy advisory. She is a member of the German Council for Sustainable Development since 2013. In 2015 she joined a CCICED task force on greening China's South-South Cooperation. She has published extensively on several topics at the interface of environment and development, including the 2030 Agenda and the SDGs, adaptation to climate change, and sustainable forest use. In the 1980s, she studied sociology at the Freie Universität Berlin where she did her doctorate in 1999 after joining DIE in 1992. Between 1999 and 2002, she was on leave and worked for German development cooperation as an environmental policy advisor in the Brazilian Amazon.

Wouter Vandenhole teaches human rights law at the University of Antwerp. He chairs the Law and Development research group at the Faculty of Law. Mr. Vandenhole graduated in law at the University of Leuven (Belgium), received an LL.M. in Law in Development from the University of Warwick (United Kingdom) and holds a PhD of the University of Leuven. He has previously held positions at the University of Leuven, Tilburg University (the Netherlands) and the European Master's Degree in Human Rights and Democratisation (Venice, Italy). His research interests include children's rights, human rights, in particular economic, social and cultural rights, and the relationship between human rights law and development. For some years now, he has focused on transnational human rights obligations, i.e. the human rights obligations of new duty-bearers, and in particular on companies.

Introduction

Markus Kaltenborn, Markus Krajewski, and Heike Kuhn

Abstract The 2030 Agenda's Sustainable Development Goals and international human rights are connected to each other in many different ways. The contributions of this volume analyse this interdependency by addressing each aspect from a more human rights-focused angle and a development-policy angle. The comparative approach underlying the contributions sheds light both on similarities and differences between these two dimensions and therefore provides a broader perspective on the relationship between development policy and international human rights protection.

More than a decade ago, the *Human Rights Quarterly* published an article with the illustrative headline "Ships passing in the night" written by *Philip Alston*, professor at New York University and meanwhile UN Special Rapporteur on extreme poverty and human rights.[1] The article described the relations—or, to be more precise—the 'lack' of relations between the human rights community and the development community: Both groups of scholars, politicians and activists work in many respects on the same issues and have similar objectives, but they hardly know of each other and therefore often act separately.

[1]Alston (2005).

M. Kaltenborn (✉)
Ruhr-University Bochum, Faculty of Law, Bochum, Germany
e-mail: markus.kaltenborn@rub.de

M. Krajewski
Friedrich-Alexander-University Erlangen-Nürnberg, Faculty of Law, Erlangen, Germany
e-mail: markus.krajewski@fau.de

H. Kuhn
Division Human Rights, Gender Equality, Inclusion of Persons with Disabilities, Federal Ministry of Economic Cooperation and Development, Bonn, Germany
e-mail: Heike.Kuhn@bmz.bund.de

M. Kaltenborn et al. (eds.), *Sustainable Development Goals and Human Rights*, Interdisciplinary Studies in Human Rights 5,
https://doi.org/10.1007/978-3-030-30469-0_1

Much of what *Alston* elaborated in his essay has not lost its topicality and validity. It is, however, doubtful whether the name-giving picture of the ships passing each other still fits today: At least since the adoption of the *2030 Agenda for Sustainable Development* in 2015 it is obvious that there are strong connections between human rights and the objectives of global development actors.[2] Already the preamble of this document explicitly states that "the 17 Sustainable Development Goals ... seek to realise the human rights of all".

Moreover, several goals and targets of the *2030 Agenda* correspond to already existing individual human rights obligations. Sustainable Development Goal (SDG) 2.1., for instance, calls on countries to "end hunger and ensure access by all people, in particular the poor and people in vulnerable situations, including infants, to safe, nutritious and sufficient food all year round" by the year 2030. This goal basically reiterates the obligations spelled out in Article 11 of the *International Covenant on Economic, Social and Cultural Rights* (ICESCR) and in *General Comment 12* adopted by the UN Committee on Economic, Social and Cultural Rights (CESCR) to specify this provision. *General Comment 15* elaborates the closely related right to water, and SDG 6 reflects the Comment's requirements for implementing this right. The new goals relating to global health (listed in SDG 3) roughly resemble the right to health that is enshrined in Article 12 ICESCR and further outlined in *General Comment 14*. SDG 4 closely relates to the right to education (Article 13 ICESCR and *General Comment 13*). The right to work is enshrined in Article 7 of the ICESCR and was further clarified 2 years ago in *General Comment 23*. In the *2030 Agenda*, it is reinforced in SDGs 4.4. and 8, which relate to employment.

Last but not least, the right to social security is of great importance. It is expressed in Article 9 of the ICESCR and is reiterated by the 2030 Agenda, particularly in the goal to "end poverty in all its forms everywhere" (SDG 1). To that end, SDG 1.3 calls on states to "implement nationally appropriate social protection systems and measures for all, including floors, and by 2030 achieve substantial coverage of the poor and the vulnerable". This goal corresponds with the right to social security, which has also been explicitly acknowledged in the *Social Protection Floors Recommendation* adopted by the International Labour Organization (ILO) a few years ago.

The complete list of overlaps between the SDGs and internationally recognised human rights is much longer. Based on a thorough analysis of the *2030 Agenda* the Danish Institute for Human Rights concluded that 156 of its 169 targets (more than 92 %) reflect human rights and basic labour standards.[3] Apart from the ICESCR, relevant human rights are defined in the *International Covenant on Civil and Political Rights* (ICCPR) and a variety of more specific human-rights conventions, like the conventions on the rights of women and children or the ILO conventions. They also

[2]On the following see Kaltenborn and Kuhn (2017); see also the Statement by the Committee on Economic, Social and Cultural Rights on "The pledge to leave no one behind: the International Covenant on Economic, Social and Cultural Rights and the 2030 Agenda for Sustainable Development", UN-Doc. E/C.12/2019/1 of 5 April 2019.

[3]Danish Institute for Human Rights (2016).

have multiple connections to the objectives of the *2030 Agenda*. To give an example on equality, a core principle of the rule of law: Equal rights for women and men to the enjoyment of all civil and political rights set forth in the ICCPR (Article 3) is a long-standing demand but has been achieved only in very few countries until today. Therefore, SDG 5 calls on achieving gender equality by ending all forms of discrimination against women and girls everywhere and ensuring women's full and effective participation and equal opportunities for leadership at all levels of decision-making in political, economic and public life. Taking on board the capacity and talents of half of the world's population could make a huge difference in solving urgent global problems.

Ensuring all these human rights is the duty of the states where the rights holders live. The *2030 Agenda* recognises this fundamental obligation too, when it states that "each country has primary responsibility for its own economic and social development". However, it is also well recognised that the global development goals will not be achieved without revitalising and expanding the Global Partnership. One implication is that official development assistance (ODA) will continue to be a prominent part of the effort. This intergovernmental aspect of the *2030 Agenda* (SDG 17) is also included in international human rights provisions. All of the above-mentioned *General Comments* refer to Article 2 para. 1 ICESCR, which obliges the state parties to take steps to realise the rights outlined. This can be done both "individually and through international assistance and co-operation, especially economic and technical".

The ICESCR duty to provide international aid is an expression of extraterritorial state obligations. Even beyond their own borders, states have an international responsibility to strive towards achieving the ICESCR goals. Hence (along with Article 56 of the UN Charter) the ICESCR is the primary hard-law basis for the Global Partnership. Although soft law still determines—at least to a certain extent—the actual scope of these support obligations as well as the thematic and country-specific allocation of duties, one can now assume that, at the latest due to the adoption of the *2030 Agenda*, the obligation to provide development aid or to engage in other forms of development cooperation does not only meet ethical requirements, but is indeed deeply rooted in international law.[4]

In the preparatory stage of the *2030 Agenda*, several UN human rights experts stressed the close link between human rights and development policy objectives: "Human rights norms and standards provide concrete guidance as to how goals and targets for the post-2015 development agenda should be framed. Governments have already committed to uphold human rights in numerous international treaties. Grounding development priorities in human rights is not only a legal and moral imperative, but can also enhance effectiveness and accountability."[5] At the same time, Navi Pillay, former High Commissioner for Human Rights, emphasized: "(P)eople across the globe are demanding that human rights be at the centre of the

---

[4]Kaltenborn and Kuhn (2017).

[5]Statement by 17 Special Procedures mandate-holders of the Human Rights Council on the Post-2015 development agenda, 21 May 2013, https://newsarchive.ohchr.org/EN/NewsEvents/Pages/DisplayNews.aspx?NewsID=13341&LangID=E.

new development agenda. Around the world, in the 88 national consultations and 11 thematic UN consultations on the new agenda, in which more than a million people have participated, this has been the most emphatic and consistent demand."[6] It seems that this demand has been heard by UN officials, politicians and diplomats when they formulated the goals and targets of the new development agenda. However, the question arises as to what significance the strong emphasis on the human rights approach has for the actual implementation of the document.[7]

The present volume takes the interconnectedness of the *2030 Agenda*'s SDGs and international human rights as the starting point for deeper analyses of the various aspects of the interplay of sustainable development and human rights. The contributions to this volume address each aspect from a more human rights-focused angle and a development-policy angle. For the most part, this is achieved by one contribution focusing more on the human rights dimension of the issue and a corresponding contribution by a scholar with a stronger emphasis on the SDG aspects. In some cases, both perspectives are combined in the same article. In both cases, the comparative approach highlights similarities and overlaps as well as differences between the two dimensions and allows to close gaps which would remain if only one perspective would be at the center of the discussions.

The first aspect addressed concerns poverty reduction. *Hans-Otto Sano* asks how a human rights-based approach can contribute to poverty reduction by focusing on the relevance of human rights in SDG 1. He critically reflects on the tendency of human rights research to imbed poverty analysis and argues that research-based evidence on human rights and poverty reduction is only modestly available and can mostly be found in local studies. *Sano* suggests that experiences from human rights-based endeavours at the local level need to be taken into account when addressing how human rights-based approaches can contribute to the implementation of SDG 1. Positive examples concern cases of empowerment processes and improved equal access to services. New technology may also offer opportunities for empowerment of the poor and for greater rights-based accountability. *Sano* concludes that the struggles of social actors from below should be recognized and be given voice, even when human rights are discussed with a global perspective.

Following the focus on poverty reduction, the volume turns next to social protection and health-related issues. *Markus Kaltenborn* assesses the human rights framework for establishing social protection floors and achieving universal health coverage and therefore connects issues of social and health protection, while *Delanyo Dovlo* discusses the right to universal health coverage and the SDGs in Africa. *Kaltenborn* takes *General Comments No. 14* and *19* of the CESCR which specify the contents of the right to health and the right to social security as a starting point. He shows that the main challenges associated with the implementation of

[6]Keynote remarks, Open Working Group on Sustainable Development Goals, 6th Session Interactive exchange on "Human rights, the right to development, global governance", 3 December 2013, https://www.ohchr.org/Documents/Issues/MDGs/SpeechASGforHC_13Dec2013.pdf.

[7]Cf. in this context also Winkler and Williams (2017) and Collins (2018).

these two human rights have been addressed in several major international policy initiatives and global partnerships. In this respect, the *2030 Agenda* contributes to the concretization of the rights to health and social security, because it expressly obliges the international community both to implement the concept of social protection floors and to ensure universal health protection. *Dovlo* in turn focusses on the right to health and the health implications of the SDGs in Sub-Saharan Africa. He shows that some countries have improved health services coverage by removing financial barriers. African countries also increased their health budgets. However, attaining health rights in Africa also requires ensuring access to quality health services, and building effective "voice" for populations to exercise their rights. While scarce resources may require rationing health services *Dovlo* points out that innovations and ICT technology can help realise access to health care for all.

As already mentioned, gender equality is a fundamental human rights principle and a cornerstone of the *2030 Agenda*. *Beate Rudolf* therefore looks at the relevance of the *Convention on the Elimination of All Forms of Discrimination Against Women* (CEDAW) for the Implementation of the SDGs while *Josephine Odera* and *Judy Mulusa* assess the prospects of the SDGs for gender equality and women's' empowerment. *Rudolf* recalls that CEDAW is the core international human rights treaty on women's equality in all fields and produced a wealth of information on causes of discrimination against women, on gaps in implementing women's human rights that prevent their full and equal participation in all areas of life as well as on successful strategies and instruments to address the structural causes of gender-based discrimination. *Rudolf* argues that CEDAW implementation can therefore be used for promoting gender-sensitive SDG implementation in light of synergies between the SDGs and CEDAW, in particular with respect to the national, regional, and global follow-up and review processes under the SDGs. Rudolf also points out that National Human Rights Institutions (NHRI) can play a crucial role in this regard. *Odera* and *Mulusa* proceed from the premise that the developmental and political goal of reducing gender inequalities remain largely unmet. They recall that the SDGs stand-alone goal on gender is more comprehensive than the earlier approach of the Millennium Development Goals (MDGs) and encompasses a potentially transformative commitment for the rights of women and girls. The authors ask if the agenda and the theory of change embedded in the SDGs are transformational for women and girls. They argue that the influence of the SDGs may be most significant in their ability to transform the idea of development from a purely economic approach to one that addresses the rights perspective. *Odera* and *Mulusa* show that gender equality is a cross cutting development issue that must be addressed in a multidimensional way for the rights of women and girls to become a reality and for the SDGs to deliver for women and girls as equal partners. They conclude by suggesting various instruments and mechanisms with the potential for transformative action.

Concerning the perspective of international labour rights and human rights of work in light of SDG 8, *Christoph Scherrer* takes a pessimistic view and argues that this goal will not be achieved. His main argument is based on the observation of the abundance of persons offering their labour power in relationship to the limited demand for their labour which stems from the insufficient absorption of peasants set free from their land. *Scherrer* shows that in many late industrialising countries most of those who are leaving agriculture do not find gainful employment even at the current junction. In fact, many of the late industrialisers are prematurely de-industrialising. In order to explain the lack of absorption capacity of industries and productive services *Scherrer* looks at demographic pressures, restrictions on migration, productivity differentials vis-à-vis the Global North and the few successful late industrialisers, constraints on the promotion of industry stemming from neoliberal globalization and challenges stemming from the colonial heritage such as the lack of societal trust.

Addressing and reducing inequality is a key human rights concern and also of great significance for the Agenda 2010. *Heike Kuhn* argues that growing inequality has a significant impact on societies and the potential to undermine democracy. Stressing that the global community has agreed for the first time ever on the goal to reduce inequality within and among countries (SDG 10), *Kuhn* provides an overview of the reduction of inequality from a legal-developmental perspective, discussing the social, economic and ecologic dimension of inequality, the reason behind the highly complex SDG 10, its genesis, the long-standing idea of international solidarity, progress reporting on this SDG, Germany's approach to implement SDG 10 and the road ahead.

The interplay and connection between the SDGs and human rights may not always be based on a liberal or progressive agenda as *Liora Lazarus* shows in her analysis of the "coercive sting" in SDG 16. *Lazarus* argues that SDG 16 is the latest symptom of the securitization of the rule of law and human rights—a process implying a shift in conception whereby human rights and the rule of law no longer embody limitations on the coercive state. Instead, these concepts are now also linked to demands for coercion, and ultimately law and order in fragile states. *Lazarus* suggests that the development movement itself may be part of this securitization process and warns of the implications of such a shift.

The relevance of environmental concerns and climate change for human rights and the SDGs is again addressed by two corresponding contributions. *Alan Boyle* looks at climate change and sustainable development from a general human rights perspective whereas *Imme Scholz* reflects on the right to development from the perspective of global environmental change and the *2030 Agenda*. *Boyle* focusses in particular on the *2015 Paris Agreement* which is intended to implement the SDGs approach to climate change. Climate change will also affect the enjoyment of human rights in many ways, but its causes and effects are too numerous and too widely spread to respond usefully to individual human rights claims. *Boyle* stresses that human rights law as a whole requires states to comply with the *Paris Agreement* and calls upon UN human rights bodies to act accordingly and hold states accountable in this respect. *Scholz* on the other hand departs from the observation that the conceptual and legal relationship between human rights, human development and environmental protection is not straightforward. The *2030 Agenda for Sustainable Development* and the *Paris Climate Agreement* adopted in 2015 link improvements in human development to human rights and to mitigating global changes in climate and the environment. The *UN Declaration on the Right to Development* (UNDRTD) of 1986, however, does not include any explicit obligation to protect the natural environment, and to contribute to the provision of global environmental goods. *Scholz* therefore takes a closer look at this declaration and how it could be linked with global environmental change. Finally, *Scholz* proposes two concepts that could help to situate the UNDRTD within the challenges of the twenty-first century as exemplified in the *2030 Agenda*. First, humanity should be introduced as a third category of right-holders (in addition to individuals and groups) and second, the rights of life forms should be established to transcend the conceptual boundaries of human rights and develop norms that govern the interdependencies between humans as well as plants and animals in the broadest sense.

The last two contributions to this volume address horizontal issues which concern all SDGs and human rights. They relate to the role of public and private actors and to extraterritorial human rights obligations. *Jens Martens* assesses public and private actors and means to implement the SDGs and calls for reclaiming public policy space for sustainable development and human rights. Departing from the observation that the *2030 Agenda* declared that public finance has to play a vital role in achieving the SDGs, he recalls that in recent decades, the combination of neoliberal ideology, corporate lobbying, business-friendly fiscal policies, tax avoidance and tax evasion has led to a massive weakening of the public sector and its ability to provide essential goods and services and to fulfill its human rights obligations. *Martens* claims that the proponents of privatization and public-private partnerships (PPPs) use these trends to present the private sector as the most efficient way to provide the necessary means for implementing the SDGs. However, as experiences by affected communities have shown, privatization and PPPs involve disproportionate risks and costs for the public sector and can even exacerbate inequalities, decrease equitable access to essential services and jeopardize the fulfilment of human rights. *Martens* concludes that it is high time to counter these trends, reclaim public policy space and take bold measures to strengthen public finance and weaken the grip of corporate power on people's lives.

In the last contribution to this volume *Wouter Vandenhole* looks beyond national borders and calls for a "division of labour" for sustainable development which builds on extraterritorial human rights obligations. He analyses the role to be played by external governmental and intergovernmental actors in bringing about sustainable development from a human rights perspective. *Vandenhole* discusses the strengths and weaknesses of the right to development and extraterritorial human rights obligations, and identifies five challenges for human rights law: the legal status of the obligations to cooperate internationally; the distributive allocation of extraterritorial obligations; the triggers of extraterritorial human rights obligations; the scope of the extraterritorial obligation to cooperate for development; and the ability of human rights law to engage with strong definitions of development, which take growth agnosticism as their starting point.

References

Alston P (2005) Ships passing in the night. The current state of the human rights and development debate seen through the lens of the millennium development goals. Hum Rights Q 25:755–829

Collins L (2018) Sustainable development goals and human rights: challenges and opportunities. In: French D, Kotzé LJ (eds) Sustainable development goals. Law, theory and implementation. Edward Elgar Publishing, Cheltenham, pp 66–90

Danish Institute for Human Rights (2016) The human rights guide to the Sustainable Development Goals, http://sdg.humanrights.dk/

Kaltenborn M, Kuhn H (2017) Acceptance, not enforcement – the legal implications of the UN-2030-Agenda and the Sustainable Development Goals. In: Development and Cooperation (D+C) April 2017, https://www.dandc.eu/en/article/2030-agenda-largely-reflects-and-reinforces-existing-international-law

Winkler IT, Williams C (2017) The Sustainable Development Goals and human rights: a critical early review. Int J Hum Rights 21(8):1023–1028

Markus Kaltenborn is Professor of Public Law at the Ruhr-University Bochum, Faculty of Law. He is Director of the Institute of Development Research and Development Policy (IEE) at Ruhr University Bochum, member of the German Institute for Human Rights and of the supervisory board of Oxfam Germany e.V; in recent years he has also been Visiting Lecturer at the Institute for Social Development of the University of the Western Cape (Cape Town). His main areas of research are Health Law, Social Protection Law, the Law of Development Cooperation and Human Rights Law. He is (together with Katja Bender and Christian Pfleiderer) editor of the book Social Protection in Developing Countries. Reforming Systems (Routledge 2013) as well as (together with Philipp Dann und Stefan Kadelbach) of the edited volume Entwicklung und Recht (Development and Law, Nomos 2014). Furthermore he is the author of a study on Social Rights and International Development (Springer 2015).

Markus Krajewski holds the Chair in Public Law and Public International Law at the University of Erlangen-Nürnberg. His research focusses on international economic law, human rights, European external relations and the law of public services. He heads the Interdisciplinary Research Centre for Human Rights Erlangen-Nürnberg (CHREN) and is one of the programme directors of the Master programme in Human Rights at the University of Erlangen-Nürnberg. He also chairs the Board of Trustees of the German Institute for Human Rights and is Secretary-General of the German Branch of the International Law Association.

Heike Kuhn is Head of Division "Human Rights, gender equality, inclusion of persons with disabilities" at the Federal Ministry for Economic Cooperation and Development (BMZ), Bonn/Berlin). She has a background of more than 25 years in European and multilateral development cooperation, human rights, financing and programming. Ms. Kuhn has long-standing experience in leading and managing multilateral negotiations at European and UN-level, lately as Executive Board Director to the International Fund of Agricultural Development and Alternate Representative at the Permanent Representation of the Federal Republic of Germany to the International Organizations in Rome/Italy. Ms. Kuhn holds a Ph.D. in Administrative Sciences from the University of Speyer (Germany) on "The social dimension of the European Community". She is a fully qualified lawyer, having studied in Marburg and Freiburg/Germany, Moscow/USSR, Los Angeles/USA and Genoa/Italy.

How Can a Human Rights-Based Approach Contribute to Poverty Reduction? The Relevance of Human Rights to Sustainable Development Goal One

Hans-Otto Sano

Contents

Abstract Addressing how a human rights-based approach can contribute to poverty reduction, the chapter reflects critically on the tendency of human rights research to imbed poverty analysis in somewhat undocumented allegations such as for instance the impact of neoliberal policies. Research based evidence on human rights and poverty reduction is only modestly available and mostly in local studies. The chapter argues that there are experiences from human rights-based endeavours at the local level that need to be taken into account when addressing how human rights-based approaches can contribute to the implementation of Sustainable Development Goal 1 (SDG 1). Positive cases of empowerment processes and improved equal access to services exist. New technology may also offer opportunities for empowerment of the poor and for greater rights-based accountability. Such examples must be included rather than an exclusive focus on negative developments, for instance, with respect to deteriorating equality at national and global levels. The struggles that social actors undertake from below should be recognized and be given voice, even when human rights are discussed with a global perspective.

H.-O. Sano (✉)
The Danish Institute for Human Rights, Copenhagen K, Denmark
e-mail: hosa@Humanrights.dk

M. Kaltenborn et al. (eds.), *Sustainable Development Goals and Human Rights*,
Interdisciplinary Studies in Human Rights 5,
https://doi.org/10.1007/978-3-030-30469-0_2

1 Introduction

Are human rights effective in reducing poverty? And in what ways are they relevant in the context of SDG 1 to ending poverty? In this chapter I aim to provide reflections on these questions. Poverty is not a concept grown in the human rights garden and human rights scholars and actors do not always manage to undertake in depth analysis, when it comes to analyses of poverty numbers, groups, and mobility. An analysis of how human rights contribute to poverty reduction will therefore prompt interdisciplinary approaches, typically a combination of a development and human rights approach in the context of the Global South.

In 2000, Mary Robinson, former High Commissioner for Human Rights described poverty as the "worst human rights problem today".[1] Since then international attention given to poverty as a significant human rights challenge has largely disappeared.[2] The lack of priority given to poverty during the last decade—and even before—has implied that the international human rights discourse has had a diffuse agenda with respect to poverty and development. While human rights scholars and activists consider poverty important as it pertains to discriminatory practices, the subject tends to be overridden by discussions of vulnerability. Knowledge about who exactly is "poor" is weak, and analytical work on changes in poverty tends to be missing. The lack of attention in this field has consequentially resulted in little evidence being accumulated on human rights and poverty reduction. Local and international human rights NGOs have continued to combine a human rights-based approach with poverty reduction, but these experiences are not always well-documented and questions remain as to their sustainability.[3] Where does this then leave the human rights community with respect to the SDG 1: End Poverty in All Its Forms Everywhere? What kind of role can human rights play in addressing the five targets of this goal? The targets address both number, access to services, and social protection systems.[4]

---

[1]Vizard (2006), p. 3.

[2]See Alston (2017b), the Special Rapporteur on Extreme Poverty in his most recent report, p. 5: *"For its part, the human rights community has had all too little to offer in response to the profound challenges associated with deep economic insecurity. The human rights to an adequate standard of living, to work and to social security have been very low on the list of human rights priorities of the major human rights groups and of the principal international and regional human rights organizations, with the exception of the International Labour Organization (ILO)."*

[3]See for instance Schmitz (2012), p. 540. See also Dugard (2014). Also Duni et al. (2009), Broberg and Sano (2017), Mustaniemi-Laakso and Sano (2017).

[4]The first two targets address the number of men, women and children living in poverty and in extreme poverty. Target 3 addresses the need for implementation of appropriate social protection systems, including floors. Target 4 relates to the equal access to resources and to services, in particular of the poor and vulnerable, including control over land and inheritance, while Target 5 addresses the need for resilience of the poor and of those in vulnerable situations. See also Sano and McInerney-Lankford (2016). In the present chapter, I shall deal with access to services, relevant for target 4, while not with access to for instance land resources. Also, the chapter will not address target 5 on resilience of the poor.

The general response of the human rights community when discussing poverty and human rights (few studies or reports have thoroughly addressed human rights and SDG One) has been formulated across three different lines of arguments. One line of argumentation takes shelter in a critique of the neoliberal agenda. This is in many ways a familiar argument dating back to the 1990s, but it is partly inadequate with respect to the scale of today's poverty challenges. Another strategy is to develop a normative argument. Thomas Pogge and his colleagues have developed one normative approach to poverty reduction, i.e., the moral obligation to eradicate poverty.[5] However, there are also new normative approaches. The Special Rapporteur on Extreme Poverty has developed thinking on the need for a universal basic income.[6] Questions remain, however, on the effectiveness of these approaches in terms of more immediate impact on groups living in poverty.

Thirdly, there are human rights scholars and actors who have made an effort to understand how human rights efforts work in various contexts, partly with respect to poverty, and based on case studies. I shall draw on some of these studies including sometimes conflicting evidence. Many studies in the human rights field are case studies inspired by human rights-based bottom up approaches. However, I shall also draw on data on national poverty reduction results in order to capture national trends of poverty.

What follows is a development of these three arguments and what then may be viewed as the most feasible strategy in determining how the human rights community can contribute to SDG 1.

2 The Critique of the Neoliberal Agenda

A repeated view among human rights scholars is that many social problems in the Global South are attributable to a neoliberal world order, or to market fundamentalism as Samuel Moyn coined the neoliberal policy trend dating back to the 1980s.[7] I define neoliberalism as a theoretically based set of policies which emphasize that value in society is determined by market-driven competition and that economic growth will be engendered by deregulation of public sector control and by allowing space for private entrepreneurship. Neoliberal policy designs stress general downsizing of the role of the public sector in the economy, enhanced competition, and market-driven change where market prices are instrumental in allocating resources in the economy.[8] These general explanations for many global ailments

---

[5]See Pogge (2007).

[6]Alston (2017a).

[7]Moyn (2014). Moyn's project is as a counterclaim to Susan Marks and Naomi Klein to dissociate the human rights revolution of the 1970s to the near simultaneous neoliberal ascendance to a mainstream orthodoxy of the 1980s. See also Moyn (2018).

[8]Birch (2017), pp. 16–31. See also Metcalf (2017).

have been forwarded over many years, but the question is if the recourse to neoliberal influence is accurate enough today to provide sufficient insight into the intricacies of how the global and the local interact. The risk of this well-rehearsed critique is that it precludes motivations to examine how market forces and political processes unfold on the ground. Not least with respect to poverty and rights interaction, empirical curiosity is warranted.

The dignitaries of neoliberalism are Friedrich Hayek and Milton Friedman, prominent libertarian scholars writing in the 1960s and 1970s, arguing for deregulation, privatization and for market reform against the dominance of the welfare state. Institutionally, neoliberal economic policies became a strong feature of policy prescriptions of the IMF and the World Bank during the 1980s and 1990s.[9] In the prescriptions promoted by the *Bretton Woods* institutions, developing countries marked by low growth, budget and balance of payment deficits and in some cases by hyperinflation were to undertake budget cuts, pursue growth- oriented policies, regulate exchange rates, and work for the privatization of sectors of the economy. These prescriptions prevailed in the so-called structural adjustment programs. Structural adjustment lending was mostly made conditional upon the alteration of specific policies such as reduction in subsidies, exchange rate adjustment according to market values of the exchange, and budget deficit cuts.[10] Easterly's skepticism on the positive impact of the structural adjustment era during the early 2000s is symptomatic of a range of critical assessments being formulated from both within and outside financial institutions. During the first decade of the 2000s the neoliberal dogma in the form of structural adjustment programs no longer had a strong appeal within the IFIs.[11] It is interesting to note that in 2013, William Easterly, a former employee of the World Bank, published *The Tyranny of Experts: Economists, Dictators, and the Forgotten Rights of the Poor*—10 years after the work quoted above. In this work, he is critical of the authoritarian technocrats who interfere unduly in the lives of the local population and who disrespect their rights.[12]

---

[9]For a broader review, see Michael Freeman in a lecture given at Dokuz Eyul University Law School, Izmir, Turkey, under the auspices of the Raoul Wallenberg Institute on 10 November 2014.

[10]For an analysis of the poverty effect of adjustment policies, see World Bank economist Easterly (2003), pp. 365 and 378–379. Easterly concludes that adjustment lending according to his analysis had no effect on economic growth, while at the same time reducing the positive impact on poverty that growth could have during expansionary periods.

[11]It is symptomatic that the themes chosen for the World Development Reports during the decade from 2001–2010 concerned poverty (2001), market institutions (2002), sustainable development (2003), services for the poor (2004), health (2005), equity (2006), youth (2007), agriculture (2008), economic geography (2009), and climate change (2010). The hard-core economic themes were abandoned to some extent in favor of social problems and sustainable development. Voices from inside the IMF wrote in 2016: "instead of delivering growth, some neolioberal policies have increased inequality, in turn jeopardizing durable expansion". See Ostry et al. (2016). See also earlier contributions by Chorev and Babb (2009), pp. 459–484.

[12]See Easterly (2013). The book emphasizes civil, political as well as economic and social rights, but does so without very strong human rights reference. Easterly's focus is in many ways inspired by Hayek, an anti-authoritarian, libertarian viewpoint.

The questions emerge then: Is there still a neoliberal project, how has it been reshaped, and why is this important to raise in a chapter on human rights, the SDGs and poverty? My points are twofold in this regard: firstly, there is a risk in human rights thinking that the normative focus will suppress a more elaborate, and sometimes needed historical and evidence-based analysis. Along with this tendency is that human rights scholars too often appeal to simplified notions of e.g. neoliberalism without acknowledging that the reality today does not accurately reflect neoliberal thinking nor does it reflect the extent to which it prevailed earlier.

Without strong reference to human rights, Heloise Weber argues that the SDG project and its goal of "Leaving No One Behind" is ideologically motivated based on neoliberal policies. The SDG provisions will privilege the upholding of commercial interests over commitments to universally ensure entitlements to address fundamental life-sustaining needs. "*The SDGs as a framework for global development integrates (comprehensively), what has been evolving as a neo-liberal development project at least since the 1980s.*" An emphasis is placed "*on realizing an unqualified conception of economic growth.*"[13]

While Weber has only scant references to human rights, her critical points on neoliberalism resonate with the recent book by Manfred Nowak, *Human Rights or Global Capitalism* from 2017. Nowak argues, "*we see the results of globalization driven by neoliberal market forces: growing inequality, poverty, and growing economic, food, financial, social and ecological crises. In addition, we witness increasing threats to our global human security resulting from transnational organized crime and terrorism, a proliferation of weapons and armed conflicts, fragile states and global climate change.*"[14] With respect to education, for instance, Nowak maintains that privatization of education is an important trend presently and that this occurs due to the influence of neoliberal policies and the structural adjustment policies of the IMF and the World Bank,[15] but his evidence of privatization is anecdotal and not based on very recent evidence (a major reference on education is from 2005); generally the reader misses clear links to either neoliberalism or to structural adjustment programs—or documentation of recent trends of privatization. Privatization of educational institutions can be motivated by other forces than neoliberal prescriptions. Furthermore, a tendency which is sometimes found in human rights studies is that they refer to other human rights studies with the result that arguments on economic and social analyses may carry the risk, in some cases, of becoming self-referential.[16]

[13]Weber (2017).

[14]Nowak (2017).

[15]Nowak (2017), p. 60.

[16]This is a tendency that is not peculiar to human rights, but the interdisciplinary perspective that is needed in discussions on economic and social trends can be found to be meager in the human rights literature. A similar tendency characterizes economic scholarly work which often completely ignores that human rights concerns are relevant in many economic domains. The most recent World Development Report of the World Bank from 2018 has education and learning as a theme: Learning to Realize Education's Promise. It contains no references to "Privatization" or to

So while a neoliberal scapegoat prevails at least in some quarters, is this an entirely misconceived proposition? With globalization and the de-linking of social relations from territorial geography, the power of market forces has not been diminished. Furthermore, the rise of income inequalities has resulted in the fact that the working and lower middle class groups in the Western world have become losers in the last 20 years of globalization, while the winners have been the poor and middle classes of the Asian nations, according to Branko Milanovic. Significantly, winners have also been the richest groups in the old Western world.[17] At the same time, it should also be emphasized that during the period from the 1990s to the present, equality of opportunity has increased as measured by access to basic education and in some cases basic health services.[18] Along with the processes of enhanced economic inequalities and with winner and loser trends of globalization, perspectives on populist and neo-conservative and nationalist influence have emerged as important descriptive and explanatory elements of poverty and exclusion.[19]

In their assessment of socio-economic rights in South Africa, Langford et al. describe South Africa as the awkward hybrid of neoliberal economic policies and a corporate welfare state.[20] They refer to neoliberalism as an emphasis on trade flows, policies of exchange rate stability, competitive labour markets, and friendly investment environments.[21] However, what is perhaps also characteristic of states such as South Africa and their response to prevailing economic downturns has been populist and corrupt policies that may have rhapsodic reference to land rights, as in Zimbabwe, or to urban jobs in South Africa, but lack real impact on poverty.[22]

In essence, the critique of a neoliberal world order today may miss a focus on neo-conservative and populist influence on policies, and they may fail to see that free markets propounded by neoliberals are replaced in part by protectionist markets favoured by conservative power holders in alliance with free market corporate sectors. An important trend is also that whereas the Bretton Woods institutions earlier could be seen as propounding the neoliberal agenda, the agenda today lacks prominent institutional agency at the international level. These broad policy trends constitute the complex patterns in which poverty and human rights policies have to

---

"Privatize", but seven references to "private sector", nine if footnotes are included. The report assesses the excluding and negative impact of school fees, see e.g. pp. 117–118.

[17]See Milanovic (2016). See also Oxfam (2017).

[18]Millennium Development Goal measurement. See for instance narrowing disparities in female primary school completion rates or in infant mortality rates.

[19]Milanovic speaks of populism and nativism, see World Development Report of the World Bank (2018), pp. 204–211.

[20]In his analysis of rights-based change in Latin America, Gledhill (2009), p. 38, argues that the forces of neo-liberalism and the multilateral institutions now seem to recognize claims to collective land from indigenous groups as long as these are not interfering with capitalist interests in land or with sub-soil resources.

[21]Langford et al. (2014).

[22]See Ismail (2015).

operate. In order to understand how this may work, it is paramount to gain evidence on what works and what is workable under existing political economy relations.

3 The Universal Basic Income

In a report to the Human Rights Council during the summer 2017, the Special Rapporteur on Extreme Poverty, Philip Alston, presented new perspectives on poverty and human rights by forwarding ideas on supplementing the existing social protection systems with a universal basic income.[23] Alston based his strongly formulated report on the profound challenges associated with deep economic insecurity. Rights to an adequate standard of living and to social security have been very low on the list of priorities of major human rights groups and of international and regional human rights organizations. He drew up an image of the human rights system that—faced with the challenges of addressing economic insecurity—will keep marching along a path mapped out long ago in a zombie mode that will gradually drain the lifeblood of the enterprise (p. 5).

The trends that need to be addressed, according to Alston, are:

- The precarious nature of unemployment;
- The difficulties of regulating the labour market;
- The redundance of part of the labour force by automation and robotization;
- The unstoppable growth in inequality; and
- The ascent of a new neoliberal agenda accompanied by fetishization of low tax rates, demonization of the administrative State, deregulation as a matter of principle, and the privatization of the remaining State responsibilities in the social sector that would undermine prospects of social rights respect.

The universal basic income would be an entitlement payable to all in society regardless of income, wealth, age, and gender. Alston recognizes that there are substantial costs involved—a calculation is presented for Catalonia in Spain and for the Democratic Republic of Congo predicated on an estimate amount of 25% of national per capita income.[24] However, the utopia should not be rejected out of hand. Policymakers at national and international levels need to develop creativity in social policy that is capable of responding to technological challenges and other developments.

However, the image drawn up by Alston that there are no promising pathways on the human rights horizon tends to miss two important perspectives: the reduction in the numbers of poor people across the Global South and the existence of human rights struggles at the local level—the experience of which should not be ignored.

[23] Alston (2017a).

[24] See the report elaborated in OECD (2017).

4 Human Rights and Poverty in Local Contexts

The available literature on human rights and development and on human rights-based approaches can help in answering three questions that are crucial to understanding the kind of challenges that will influence SDG implementation of targets and indicators of SDG 1.

- How entrenched is poverty both in terms of numbers and in terms of social mobility?
- How is equal access to resources and services hampered by existing institutional structures?
- How can rights-holders in turn forward effective claims against the duty-bearers?

Poverty in Numbers and Processes of Change

In most countries the share of population below a poverty line defined by e.g. incomes or below standards of deprivation set by health, education and living standards indicators as the ones defined in the Multidimensional Poverty Index (used in this chapter) has fallen significantly. Table 1 compares the incidence of national, urban, and rural poverty during the second half of the former decade (Ghana is an exception as its timespan dates back to 2003).

Table 1 The incidence of poverty

Tanzania	2015/16	2010	2008
National	56.6	61.1	65.7
Urban	27.7	34.8	39.5
Rural	68.6	68.9	72.8
Ghana	2014	2008	2003
National	33.7	41.9	58.7
Urban	17.6	20.6	36.2
Rural	49.4	58.4	74.9
South Africa	2014/15	2012	2008
National	9.2	10.5	17.8
Urban	3.6	5.4	9.4
Rural	17.2	18.3	30.4
Bangladesh	2014	2011	2007
National	41.3	49.6	59.1
Urban	22.5	26.1	37.4
Rural	48.2	57.0	65.9
Indonesia	2012	2007	N/A
National	15.5	20.8	
Urban	10.2	13.2	
Rural	20.7	26.4	

Five country cases during the present and the former decade
Source: Oxford Poverty Human Development Initiative. *Multidimensional Poverty Index*

What is noteworthy is the variation in the incidence of poverty. More than two-thirds of the population in rural Tanzania are affected, while only about 4% of the urban population in South Africa are. In all five included country cases, poverty has fallen, with the exception of Ghana and Bangladesh where the fall is steady between the years of the table, and the fall in percentage points is most substantial during the former decade. In South Africa, the decrease is minimal between 2012 and 2014/15, i.e. during the recent period of Zuma's populist regime.

The methodology for identification of poor households devised under the Oxford Poverty and Human Development Initiative (OPHI), is the so-called Global Multidimensional Poverty Index (MPI). The MPI measures overlapping and simultaneous dimensions of deprivations. Similar to the Human Development Index, it measures indicators across three dimensions: health, education and living standards. The effort to develop a multidimensional poverty index was partly inspired by basic needs thinking. Household incomes or expenditures are not part of the measurement.[25] The MPI is not based on a rights-based method, but it is premised on measuring deprivation that relates to the right to education, the right to health and to food, and to adequate living standards.[26] However, it must be underlined that the indicators used in the index have not been conceived with a human rights-based approach in mind.

The patterns in Table 1 indicate therefore that poverty is diminishing in countries of the Global South, but with a marked slowing of pace in the rural areas. The mobility of people moving out of poverty is significant during the observed decades, something that may bode well for the ending of poverty; however, the challenges of entrenched poverty prevail in rural areas in particular. If human rights work should contribute to the realization of the SDG 1, part of the work should therefore take place in the rural areas or should have an impact there.

Equal Access to Services and Institutional Impediments

When raising questions on equality of access to services, the literature on the implementation of a human rights-based approach offers insights that are based on evidence, mostly qualitative, and by evaluation studies. These studies are valuable, but rare, and even rarer is the availability of studies based on research methods: most of the insights that can be obtained from these works are based on experiences from project-based work. This means that the reality outside the financed project reviewed is not taken into account in a detailed manner. Another limitation of the available studies is that there is no documented focus on poverty. Human rights programming often follows a logic of disaggregation which pays attention to discrimination

[25]For more see Alkire and Robles (2015).

[26]Relevantly, we could also raise questions on unequal access to resources, e.g. women's land and inheritance rights, but this issue is not addressed in this chapter.

against women, maybe against elderly groups, and according to ethnic groups, but it is not very attentive to how poverty is redressed through human rights-based programming. In brief the question is this: can the implementation of human rights-based approaches contribute to the redress of prevailing service inequalities in favour of poor households? What are the obstacles?

Three studies[27,28,29] point to gaps between the rhetorical ambition of the human rights-based approach and actual implementation. Destrooper emphasizes gaps between the discursive attention to e.g. human rights principles[30] and what is actually operationalized in programming, and subsequently realized or implemented. The program which she and her team reviewed is a UNICEF program on water and sanitation. However, the research expresses a negative assessment according to all 12 scores reviewed. In terms of equality, non-discrimination and inclusion, the researcher concludes that the instruments developed for achieving inclusivity—such as quotas—are often not imbedded in the local reality (p. 805). Thus, this research documentation raises questions on the realism of particular elements of a human rights-based approach, but also questions the consistency with which the approach was implemented in this particular program.[31] The appropriateness of the rights-based approach in the form of the UN defined principles and modalities could therefore be called into question in the context of this particular work.

In Ghana, Sano and Anyidoho reviewed access to education, health, water and sanitation services in two villages in the Volta Region, one having received support from Danida and Government of Ghana funding. Their report tells a more positive story in terms of access to services by the poor in the village supported, but it also points to the substantial differences of access between this favoured village and the one in the neighbouring district. The differential approach between villages and districts is due in part to the government's policies which favour effective districts (measured according to governance indicators) over less effective ones. Thus, a conflict between governance and human rights policies may result from the fact that both policy sets have an influence. This conflict can be said to reflect exactly a conflict between a human rights-based approach and an effectiveness and efficiency perspective more aligned to neoliberal thinking. Concerning the access to services, the quantitative data of the researchers showed improvements in local perception of water services, and perceptions of improvements among the poor and non-poor groups with respect to sanitation in the village supported by Government. In the

[27]Destrooper (2016).

[28]Sano and Anyidoho (2016).

[29]Critical Rights and Gender Consult (2017).

[30]See for instance OHCHR (2018).

[31]The article is, however, not very informative on the actual results of the program. This author had to identify an evaluation that was actually undertaken of the program in 2012 to get an idea about that there were some positive results. EAA pour la compte du gouvernement de la République Démocratique du Congo et Le Fonds de Nations Unies pour l'Enfance (UNICEF), 2012. Evaluation du Programme Ecole et Village Assainis. Rapport Final.

contiguous village, only a minority saw improvements in sanitation services. With respect to education, more than 90% of the households had on average three to four family members attending primary school, and with no marked differences between the two villages. In the less affluent village, quite a lot of the parents indicated that they paid for education.

With respect to malnourishment and mortality, more of the poor households had members who had experienced child mortality. Nearly all households in both villages attended the village clinic and used the town hospital. In both villages, about three-quarters of the population perceived that health services had improved. This could be due to that fact that services are subsidized under the National Health Insurance Scheme. The report went on to conclude that the human rights-based approach, where operational locally, contributed in making access to services more equitable. Thus under-five malnourishment was significantly lower in the community which had benefitted from human rights-based support.[32]

With respect to the DanChurchAid documentation of the results of the rights-based approach, the consultant (Sarah Forti) analyzed the results in terms of three concepts: *Representation*, *Recognition* and *Redistribution*. She reviewed supported projects in Cambodia and Uganda.

Regarding *Representation*, across the communities in Cambodia, there were testimonies of strengthened qualitative and quantitative participation into claiming land rights. In Uganda, she observed increased participation of marginalized communities through community monitors. With respect to *Recognition*, she found indications among duty-bearers in Cambodia that they recognized that rights-holders could make land rights claims and that there were legitimate claims in terms of land evictions (p. 25). In Uganda, the community monitors were recognized by the rights-holders as well as—in some cases—by duty-bearers (p. 24). Concerning *Redistribution*, reallocation of grabbed land had occurred in Cambodia and there were signs of slight shifts in the balance of power between rights-holders and duty-bearers. In Uganda, limited reallocation of resources could be seen especially when women were chairing committees; however, this happened in situations with decreasing budgets.

The study therefore found that firstly the main effects of the human rights-based approach were at the individual level in terms of accessing or re-accessing livelihood resources and public services rather than at the systemic and institutional level; secondly that *Representation* seemed to be the aspect that could best be fulfilled from the rights-holder perspective (p. 6). Across cases it was possible to address human rights violations in a better way, i.e. a matter of *Recognition*. However, *Redistribution* seemed difficult to strengthen solely from the point of view of rights-holders.

After reviewing these studies—all addressing the matter of services and equality from different angles—it seems plausible that a human rights-based approach

[32]Sano and Anyidoho (2016), pp. 44–45.

does have the potential of redressing some inequalities of service access.[33] The study from Ghana points to the results achieved in service access and quality in one community benefitting from human rights-based support. The DanChurchAid study reveals limited results in redistribution of resources, but also underlines that "a clear value added of HRBA is its effects in attempting to re-establish social justice. . . ."[34]

Agency, Advocacy, and Voice

In this chapter, no distinctions are made between the concepts of agency, advocacy, and voice simply because irrespective of the origins of these concepts they all relate to the human rights-based perspective of rights-holder demands being addressed to duty-bearers. A human rights-based approach presupposes that rights-holders are capable or are enabled and empowered to raise claims of rights fulfilment against duty-bearers. A human rights-based approach therefore envisages citizens as drivers of change. However, in many of the studies that address poverty and rights, rights-based agency takes place through local civil society groups or through intermediaries such as the community monitors mentioned in the study from Uganda above.

In their study from Cameroon, Duni et al. write about a paralegal program in the Mbororo community that facilitated a move from marginality and exploitative patron-client relations toward a "negotiated clientilism", i.e. not the realization of ideal forms of participatory citizenship, but nevertheless a significant change.[35] In the same volume, Gledhill argues that it is not realistic to imagine that spontaneous bottom-up action of poorer citizens will provide significant change in a neoliberal order, but at the same time he argues there is everything to be said for measures that enhance the capacity to organize—not only to make demands, but also for these citizens to take greater control in the production of their identities and enhance public understanding of their lives and problems.[36]

---

[33]This is also the conclusion arrived at by Schmitz (2012). Schmitz reviews the local community work of Plan International: Schmitz reports results in the effectiveness of service delivery, in rights awareness and ownership, but not in addressing the "root causes", whatever this means, of inequality, see pp. 523 and 534.

[34]Critical Rights and Gender Consult (2017), p. 8. The importance of social justice as a feature of the human rights-based approach was also underlined by Hickey and Mitlin (2009) who contrast a human rights-based approach with Amartya Sen's perception of rights reinforcement as a project of freedom with one of social justice: "Here we would differ from Sen by suggesting that rights can be most progressively linked to a broader project of social justice rather than of freedom.", p. 227.

[35]Duni et al. (2009), p. 63.

[36]Gledhill (2009), pp. 43–44. Schmitz (2012) is also cautious in allocating change potential to rights-holders beyond the organizations representing them.

Two sets of learnings emanate from the various studies of rights implementation in the local context. First, in the context of endemic poverty, mobilization for rights always poses a challenge.[37] Organizational capacities are weak, local organizations fragmented or non-existent, and marginalized groups are without respect or self-respect, and in many cases located remotely. In Sano and Anyidoho's study, significant differences could be observed in the propensity of poor groups to complain compared to the non-poor ones.[38] In Destrooper's study from the Congo, a large gap exists between the empowerment rhetoric, the empowerment instrumentation, and the real implementation of empowerment.[39] What emerges clearly, however, is this: Empowerment strategies must therefore be realistic and contextual.

Why then discuss the perspective of agency and advocacy in a chapter dealing with rights and the SDGs? Undeniably, participation and accountability are integral concepts of a human rights-based approach—and participation efforts are mostly accompanied by efforts of social mobilization and empowerment. However, while empowerment processes to date do not seem very promising, new technologies may strengthen the prospects of advocacy and alter the conditions of participation. The advocacy ambitions of the human rights movement already combine with the social accountability efforts of governance actors,[40] but both of these efforts addressing participation, social monitoring and social mobilization may increasingly be influenced by the proliferation of the internet and by information and communication technology (ICT).

A research project, "Making all Voices Count", undertaken by the Institute of Development Studies in Sussex, UK, has just issued reports on the findings on appropriating technology for accountability. Some of the key findings are:

- Not all voices can be expressed via technology as only half of the world's population is online.
- Technologies can play decisive roles in improving services. Cheaper, quicker, simpler generation of real-time data by governments and by citizens can help target resources, resolve distributive conflicts and allow better decisions in the provision of public goods.
- Transparency, information or open data are not sufficient to generate accountability.
- Technologies can support social mobilization and collective action by connecting citizens.

[37]This is not to argue against Simmons (2009), but rather to underscore that what is also prevalent in her study that mobilization necessitates organizational strength.

[38]These differences occurred especially with respect to water, sanitation, and primary education in the poorest village. See Sano and Anyidoho (2016), p. 44.

[39]Destrooper (2016), pp. 807 and 811.

[40]Social accountability or demand-led governance is a concept used in the governance field. I have earlier written on the parallels and overlaps between social accountability and human rights, see: Sano (2015).

- Technologies can create new spaces for engagement between citizens and the state.
- Technologies can help empower citizens and strengthen their agency for engagement.
- The kinds of democratic deliberation needed to challenge systemic lack of accountability are rarely well-supported by technologies.
- Technologies alone do not foster the trusting relationship needed between governments and citizens, and within each group of actors.
- The capacities needed to transform governance relationships are developed offline, and in social and political processes, rather than by technologies.
- Technologies cannot overturn the social norms that underpin many accountability gaps.
- A deepening digital divide risks compounding existing exclusions.[41]

From a human rights perspective, and in relation to future SDG implementation, this research is valuable. The research puts some sobering conclusions to the tech optimists by pointing to that fact that new tech-enabled norms of self-service, self-help and crowdsourcing sit alongside the ascendance of the transnational tech giants that own infrastructures, algorithms and data on which e-governance work depends (p. 25). Yet the report also indicates new avenues of empowerment, participation and social mobilization. The report does not have in- depth reflections on how this may affect poorer groups, but it contains elements that can also influence these groups positively over time.

5 Conclusion

This chapter has examined how human rights and human rights-based approaches can contribute to poverty reduction, to social floors, and to equal access to services.

The main message of the analysis is that rather than being paralyzed by ominous paradigms of neoliberalism and by important trends of economic inequality within countries—which *is* an important factor—human rights scholarship could examine the evidence for positive change which also prevails in tandem with negative scenarios: in terms of reduction in poverty numbers and in terms of positive change in access to services brought about, inter alia, by human rights-based programming and by struggles for social justice. There is a need to learn from these experiences and to investigate how human rights and SDG efforts can join forces, also locally. The fact that many human rights debates are undertaken with a perspective on global, international, or national dimensions should not preclude an ambition or a curiosity as achievements or shortcomings locally.

[41]See McGee et al. (2018). See also Herringshaw (2018).

The social floor dimension, while not explicitly addressed here, has been referenced by way of the incipient discussions on universal basic income. As a vision it should not be disregarded. It is unfortunate, however, that human rights work in this context has been presented as that of being on a zombie path that would lead nowhere in terms of addressing economic insecurity. Such a pessimistic perspective does not adequately do justice to the struggles and organizational efforts that do take place locally where human rights-based approaches are being implemented, often under difficult circumstances.

In the quest to meet SDG 1, there is a call for more evidence-based work in the human rights domain, both qualitative and quantitative. Fulfilling this call may modify some of the prevailing pessimism by pointing constructively to avenues of collective action, participation and social mobilization that may certainly hold relevance and practicality to poorer groups in the context of ending poverty during the period leading up to 2030 when the present targets of the Sustainable Development Goals should be realized.

References

Alkire S, Robles G (2015) Multidimensional poverty index 2015: brief methodological note and results. Oxford Poverty and Human Development Initiative (OPHI), Oxford Department of International Development, University of Oxford

Alston P (2017a) Report of the Special Rapporteur on Extreme Poverty and Human Rights. A/HRC/35/26

Alston P (2017b) The Special Rapporteur on Extreme Poverty in his most recent report. Report of the Special Rapporteur on Extreme Poverty and Human Rights

Birch K (2017) A research agenda for neoliberalism. Edward Elgar Publishers, Cheltenham

Broberg M, Sano HO (2017) Strength and weaknesses in a human rights-based approach to international development. Int J Human Rights. https://doi.org/10.1080/13642987.2017.1408591

Chorev N, Babb S (2009) The crisis of neoliberalism and the future of international institutions: a comparison of the IMF and the WTO. Ther Soc 38:459–484

Critical Rights and Gender Consult (2017) Documenting the effects of HRBA in development. Synthesis Report. Evaluation study for DanChurchAid

Destrooper T (2016) Linking discourse and practice. The human rights-based approach to development in the Village Assaini Program in the Kongo Central. Hum Rights Q 38(3):787–813

Dugard J (2014) Urban basic services, rights reality, and resistance. In: Langford M, Cousins B, Dugard J, Madlingozi T (eds) Socio-economic rights in South Africa. Symbols or substance. Cambridge University Press, Cambridge, pp 304–306

Duni J, Fon R, Hickey S, Salihu N (2009) Exploring a political approach to rights-based development in North West Cameroon: from rights and marginality to citizenship and justice. In: Hickey S, Mitlin D (eds) Rights-based approaches to development. Exploring the potentials and pitfalls. Kumarian Press, Sterling, pp 63–64

Easterly W (2003) IMF and World Bank structural adjustment programs and poverty. In: Dooley MP, Frankel JA (eds) Managing currency crises in emerging markets. University of Chicago Press, Chicago, pp 361–391

Easterly W (2013) The tyranny of experts. Economists, dictators, and the forgotten rights of the poor. Basic Books, New York

Gledhill J (2009) The rights of the rich versus the rights of the poor. In: Hickey S, Mitlin D (eds) Rights-based approaches to development. Exploring the potentials and pitfalls. Kumarian Press, Sterling, pp 31–46

Herringshaw V (2018) Increasing citizen voice and government responsiveness: what does success really look like, and who decides?. Institute of Development Studies, Research Report

Hickey S, Mitlin D (eds) (2009) Rights-based approaches to development. Exploring the potentials and pitfalls. Kumarian Press, Sterling, VA

Ismail Z (2015) The rise of populism in Southern Africa's dominant party states. OXPOL. The Oxford University Politics Blog

Langford M, Cousins B, Dugard J, Madlingozi T (2014) Socio-economic rights in South Africa. Symbols or substance?. Cambridge University Press, Cambridge, pp 448–450

McGee R, Edwards D, Anderson C, Hudson, H, Feruglio F (2018) Appropriating technology for accountability: messages from making voices count. Institute of Development Studies, Research Report

Metcalf S (2017) The Idea that swallowed the World. The Guardian, 18 August 2017

Milanovic B (2016) Global inequality. A new approach for the age of globalization. Belknap and Harvard University Press, Cambridge, pp 18–22

Moyn S (2014) A powerless companion: human rights in the age of neoliberalism. Law Contemp Probl 77:147–169

Moyn S (2018) Human rights are not enough. The Nation, 16 March 2018

Mustaniemi-Laakso M, Sano HO (eds) (2017) Human rights-based change. The institutionalization of economic and social rights. Routledge, London

Nowak M (2017) Human rights or global capitalism. The limits of privatization, vol 1. University of Pennsylvania Press, Philadelphia

OECD (2017) Basic income as a policy option. Available at https://www.oecd.org/els/emp/Basic-Income-Policy-Option-2017.pdf. Accessed 18 Apr 2018

OHCHR (2018) What is a human rights based approach. Available at http://hrbaportal.org/faq/what-is-a-human-rights-based-approach. Accessed 18 Apr 2018

Ostry JD, Loungani P, Fuceri D (2016) Neo-liberalism: oversold? Finance Dev 52(2):38–41

Oxfam (2017) An Economy for the 99 Percent. Just 8 men own same wealth as half the world. Oxfam Press Release, 16 January 2017. Available at https://www.oxfam.org/en/pressroom/pressreleases/2017-01-16/just-8-men-own-same-wealth-half-world. Accessed 18 Apr 2018

Pogge T (ed) (2007) Freedom from poverty as a human right. Who owes what to whom? Oxford University Press, Oxford

Sano HO (2015) Social accountability in the World Bank. How does it overlap with human rights? In: LaDawn H, Stryker R (eds) Closing the rights gap. From human rights to social transformations. University of California Press, Oakland, pp 219–236

Sano HO, Anyidoho NA (2016) The right to services and poverty. A case study from the Volta Region in Ghana. Matters of concern. Human Rights Research Papers, the Danish Institute for Human Rights

Sano HO, McInerney-Lankford S (2016) Human rights and the post-2015 sustainable development goals. Reflections on challenges and opportunities. World Bank Legal Rev 7:165–183

Schmitz HP (2012) A human rights-based approach (HRBA) in practice: evaluating NGO development efforts. Polity 44(4):423–541

Simmons B (2009) Mobilizing for human rights. International law in domestic politics. Cambridge University Press, Cambridge

Vizard P (2006) Poverty and human rights. Sen's capability perspective explored. Oxford University Press, Oxford

Weber H (2017) Politics of leaving no one behind. Contesting the 2030 sustainable development goals. Globalization 14(3):400

World Bank (2018) World development report of the World Bank

Hans-Otto Sano is Director of Research at the Danish Institute for Human Rights and currently involved in three research projects at the Danish Institute for Human Rights: one on human rights-based approaches and poverty reduction, one on human rights and economics, and one on human rights research methods including how institutional theory can be applied in human rights research. Mr. Sano has worked earlier on subjects relating to governance and social accountability, on human rights indicators, and on the meaning and value added of human rights-based approaches. He has research experience from several countries in sub-Saharan Africa and from South Asia.

The Human Rights Framework for Establishing Social Protection Floors and Achieving Universal Health Coverage

Markus Kaltenborn

Contents

Abstract In its General Comments No. 14 and 19, the Committee on Economic, Social and Cultural Rights has specified the contents of the right to health and the right to social security. The main challenges associated with the implementation of these two human rights have been addressed in several major international policy initiatives and global partnerships: The 2030 Agenda now makes an important contribution to the concretization of the rights to health and social security, because it expressly obliges the international community both to implement the concept of social protection floors and to ensure universal health protection. The extra-territorial obligations deriving from the two human rights are also taken up by the 2030 Agenda.

M. Kaltenborn (✉)
Ruhr-University Bochum, Faculty of Law, Bochum, Germany
e-mail: markus.kaltenborn@rub.de

M. Kaltenborn et al. (eds.), *Sustainable Development Goals and Human Rights*, Interdisciplinary Studies in Human Rights 5,
https://doi.org/10.1007/978-3-030-30469-0_3

1 Introduction

Whenever the media or human rights literature refer to "worst attacks on human rights"[1] or "gross and systematic human rights violations",[2] they usually refer to killings, torture, mass rape, or arbitrary arrests for which states are responsible. It is not surprising that such grave violations of human rights attract the attention of the general public and also have a prominent role in the work of the major human rights NGOs. However, it should not be forgotten that there are other human rights that are to a far greater extent impaired on a daily basis. If the absolute number of people currently affected by human rights violations is taken as a benchmark, it is the rights to health and social security where the gap between what is legally required and what is actually implemented in practice is particularly wide. The International Labour Organization (ILO) assumes that 71% of the global population are not covered by comprehensive social security systems. Only 21.8% of the world's unemployed are entitled to unemployment benefits; only 68% of all people at retirement age receive regular (however mostly only minor) benefits.[3] According to the latest World Health Organization (WHO) estimates, less than half of the world's population has access to full basic medical care; every year, more than 100 million people are driven into poverty because they have to pay health services out of their own pockets.[4] Essential medicines are still inaccessible to a majority of the global population; many, often fatal, diseases could be avoided by adequate health care.[5]

Therefore, the human rights to social security and health—both enshrined in the *Universal Declaration of Human Rights* and in the *International Covenant for Economic, Social and Cultural Rights* (ICESCR)[6] as well as in several other global and regional treaties[7]—are still a long way from being implemented worldwide. At

[1]http://www.independent.co.uk/news/world/politics/amnesty-international-reveals-the-10-worst-attacks-on-human-rights-across-the-world-last-year-a6892911.html (last accessed 27 March 2019).

[2]Damrosch (2011); cf. also Human Rights Watch (2018), pp. 155, 182, 273.

[3]International Labour Office (2017b), pp. xxix–xxxii.

[4]https://www.who.int/en/news-room/fact-sheets/detail/universal-health-coverage-(uhc) (last accessed 27 March 2019).

[5]A good overview of the state of global health care is provided by WHO (2015); WHO (2018), p. 4 et seq.

[6]Art. 9, 12 *International Covenant for Economic, Social and Cultural Rights* (ICESCR), 993 UNTS 3.

[7]Art. 5e iv *International Convention on the Elimination of All Forms of Racial Discrimination* (1966, 660 UNTS 195); Art. 11 Abs. 1 e, f, 12 *Convention on the Elimination of All Forms of Discrimination against Women* (1979, 1249 UNTS 13); Art. 24, 26 *Convention on the Rights of the Child* (1989, 1577 UNTS 3); Art. 27 *International Convention on the Protection of the Rights of All Migrant Workers and Members of Their Families* (1990, 2220 UNTS 3); Art. 25, 28 Abs. 2 *Convention on the Rights of Persons with Disabilities* (2006, 2515 UNTS 3); Art. 11, 12, 13 *European Social Charter* (1951, ETS No. 005); Art. 9, 10 *Additional Protocol of San Salvador (to the American Convention on Human Rights) on economic, social and cultural rights* (1988, OASTS 69); Art. 16 of the *African Charter on Human and Peoples' Rights* (1981, OAU Doc. CAB/LEG/67/3 rev. 5, 21 I.L.M. 58 [1982]), Art. 13 f *Protocol to the African Charter on Human and Peoples' Rights on the Rights of Women in Africa* (2003, OAU Doc. CAB/LEG/66.6); Art. 36

least, there has been considerable progress in the concretisation of content of the two human rights in recent years. The relevant UN human rights bodies as well as the ILO and the WHO have provided valuable clarification work here (see Sects. 2 and 3 below). Moreover, the *2030 Agenda for Sustainable Development* has given considerable political support both to social protection and to the health care sector (see Sect. 4), which is of great significance for the further implementation of these human rights, in particular in middle- and low-income countries.[8]

2 General Comments No. 14 and 19 of the Committee on Economic, Social and Cultural Rights

At a global level, issues related to the establishment of social protection and health systems have long been debated particularly in development policy contexts, but recently these issues have also been discussed more intensively at the level of human rights. This is partly due to the fact that in the meantime much more attention has been paid to social rights than about twenty years ago. It is hardly contested in jurisprudence that these rights—and thus also the rights to social security and to health—are basically just as legally enforceable as civil rights.[9] In many legal systems (not least in the Global South) the courts are intensively involved in the implementation of international and constitutional guarantees in the area of social rights.[10] This growing importance is further underlined by the *Optional Protocol to the ICESCR*[11] which came into effect in 2013 and provides for a right of appeal for individuals and non-governmental organizations.

The Committee on Economic, Social and Cultural Rights (CESCR) convened by the UN Economic and Social Council (ECOSOC) issued two comments in the years

---

Arab Charter of Human Rights (2004; 12 Int'l Human Rights Reports 893 [2005]).—At least just as important for the legal specification of the right to social security are the standard setting activities of the ILO. In several conventions the ILO has formulated minimum requirements for various social security branches, which are today the standard for many welfare systems in the world; see ILO (2017a) and also Rodgers et al. (2009), p. 139. Probably the most well-known of these treaties is the *Social Security (Minimum Standards) Convention* of 1952 (210 UNTS 131). Until today, 58 ILO member states have ratified this Convention—which means that they have committed themselves to providing social protection in the event of illness, unemployment, old age, occupational accidents or diseases and in the case of maternity to a specific percentage of the population; survivors and family benefits are also provided for in the Convention. However, governments are allowed, by virtue of flexibility clauses, to limit their implementing obligations to individual social security branches.

[8] The following text is mainly based on Kaltenborn (2017a) and Kaltenborn and Troeppner (2017).

[9] See e.g. Coomans (2008) and Langford (2008).

[10] See the references in Kaltenborn (2017b), p. 259 et seq.; for an analysis of rights-based social policies in South and Southeast Asia see Koehler (2017).

[11] UN Doc. A/RES/63/117.

2000 and 2008—*General Comments No. 14* and *No. 19*—which define what is concretely meant by "health" and "social security" and moreover describe the governmental obligations associated with these rights.[12] Although the *General Comments* of the CESCR are soft law and thus not legally binding under international law,[13] they can be considered as the main source for the interpretation of the Covenant (inter alia as part of the monitoring work of the Committee). In legal literature on social rights they regularly form the starting point for further analysis.[14] Their outstanding position in human rights doctrine is in particular due to the fact that the representatives in the Committee are, on the one hand, independent human rights experts, but on the other hand also gain political and legal legitimation through the selection process—they are appointed to the Committee by the state parties for four years (with the possibility of re-election).

The Right to Health

According to *General Comment No. 14* ICESCR member states—meanwhile 169 in number—are required to fully respect the right to health in all their activities, to protect it from impairments by third parties (e.g. individuals and business enterprises) and last but not least to guarantee the actual conditions for its implementation ("obligations to respect, to protect and to fulfill").[15] The scope of protection set out in Art. 12 ICESCR is very broad.[16] Under Article 12 (1) ICESCR, Member States recognize 'the right of everyone to the enjoyment of the highest attainable standard of physical and mental health'; Article 12 (2) ICESCR sets out a list of those measures which are particularly important in order to realize this right (reduction of infant mortality, environmental and industrial hygiene, control of epidemics and occupational diseases, provision of medical facilities and medical care of everyone). When implementing the right to health, attention must always be paid to the socio-economic context of this human right—mainly factors such as origin, socialization, life, work and age, all of which have an impact on the health of the individual.[17]

---

[12]Committee on Economic, Social and Cultural Rights (CESCR), *General Comment No. 14 (2000) on the right to the highest attainable standard of health (Art. 12 ICESCR)*, UN-Doc. E/C.12/2000/4; *General Comment No. 19 (2008) on the right to social security (Art. 9 ICESCR)*, UN Doc. E/C.12/GC/19.

[13]See on the importance of soft law instruments both for the development of the human rights architecture and also for the implementation of human rights Forsythe (2018), pp. 15 et seq.; cf. also Lagoutte (2016). See also on the role of 'soft governance' in global social policy from a political-economic perspective see Bender (2016) and Bender et al. (2014).

[14]Cf. Shelton (2014), p. 56; Ssenyonjo (2016), p. 42; Goldblatt (2016), p. 90.

[15]*General Comment No. 14*, para. 33.

[16]Gostin (2014), pp. 20, 251.

[17]*General Comment No. 14*, para. 9; cf. also Braveman (2010) and Chapman (2010).

One of the key obligations arising from Art. 12 ICESCR is the establishment and further enhancement of national health systems.[18] Firstly, States Parties are required to provide facilities, goods and services of public health as well as appropriate government programs which ensure that the basic elements for the maintenance of public health are in place (availability). These include, but are not limited to, the provision of hospitals and other health care facilities as well as adequate medication.[19] In addition, accessibility of the relevant health care facilities, goods and services must be ensured for each person on a non-discriminatory basis, not only physically—problems can arise in particular in rural areas where people have to put up with long distances to health services—, but also in financial terms (affordability). Another important requirement is the acceptability of health programs: Healthcare should be provided in in accordance with medical ethics and cultural habits; the latter aspect in particular has often led the Committee, when reviewing country reports, to elaborate on specific practices, such as the use of traditional medicines and therapies preferred by indigenous peoples.[20] Healthcare must also be of adequate quality: This means that only medically trained staff should be employed and only scientifically tested drugs should be available. Finally, the right to health must also be enforceable for the individual, i.e. he or she must have access to effective judicial or other appropriate remedies in the event of a potential infringement of this right (accountability).[21]

The Right to Social Security

In its *General Comment No. 19* the CESCR first of all clarifies what exactly is meant by "social security". Key areas of a social security system are (similar to *ILO Convention No. 102*) the sectors of health, social benefits for older persons, protection in case of unemployment, employment injuries and occupational diseases, family and maternity benefits, and support for the disabled, survivors and orphans.[22] Social protection schemes covering these life-cycle risks have to be generally available and must be designed in compliance with the principles of human dignity and non-discrimination; in addition, access to social protection schemes must be

[18]See Krennerich (2015), pp. 24 et seq.—For a global survey of recent health policy reforms see Kuhlmann et al. (2015), pp. 135 et seq.; on the influence of global health actors on the development of national health care systems see Kaasch (2015).

[19]The Committee refers to the *WHO Model List of Essential Drugs*, http://www.who.int/medicines/publications/essentialmedicines/en/ (last accessed 27 March 2019).

[20]See the example of Venezuela CESCR, E/C.12/VEN/3, 20.11.2013, Rn. 52, 58, 66; cf. also *General Comment No. 14*, para. 27.

[21]*General Comment No. 14*, para. 59 et seq.; see in this context Gloppen (2008); Yamin (2008); Yamin and Gloppen (2011); cf. also Ferguson (2017) who gives an overview of some mechanisms through which countries' compliance with health-related human rights can be assessed.

[22]*General Comment No. 19*, para. 12–21.

guaranteed for all people in the country, which in particular has consequences for the affordability of social benefits or insurance contributions.[23]

As far as organizational matters are concerned, *General Comment No. 19* leaves the governments a fairly extensive scope of action. They may implement contribution-based social protection systems (following the concept of Bismarck's social insurance legislation) as well as tax-financed programs (e.g. social assistance, public employment programs); but also privately-run schemes or self-help measures (community-based or mutual schemes) can be elements of national social security systems.[24] As a rule, the right to social security is implemented by governmental or state affiliated (semi-public/self-governing) institutions. Insofar as the government does not itself provide social protection in a particular sector, but relies on the services of third parties (e.g. private health or pension insurance companies), it must take appropriate measures to ensure that there are no undue disadvantages for specific groups of the population through restrictions on access to services (this is part of the so-called "obligation to protect").[25] The "obligation to fulfill" requires states parties to recognize the right to social security within their national political and legal system (preferably on a statutory basis), to elaborate a social protection strategy including a plan of action, moreover to establish appropriate protection programs and to provide the population with adequate information about these programs—this obligation is particularly important in view of the large numbers of people living in remote rural areas in the countries of the Global South. Another important aspect of this type of obligation is the provision of social assistance and social services especially to poorer and disadvantaged groups of the population (if necessary on a non-contributory basis).[26]

Progressive Realization, Core Obligations and International Assistance

In *General Comment No. 14*, the members of the CESCR state that

> for millions of people throughout the world, the full enjoyment of the right to health still remains a distant goal. Moreover, in many cases, especially for those living in poverty, this goal is becoming increasingly remote. The Committee recognizes the formidable structural and other obstacles resulting from international and other factors beyond the control of States that impede the full realization of article 12 in many States parties.[27]

Indeed, the claims associated with the broad scope of Art. 12 ICESR, at least at first sight, appear scarcely realistic since even the richer members of the international

[23] *General Comment No. 19*, para. 11, 22–27.
[24] *General Comment No. 19*, para. 4–5.
[25] *General Comment No. 19*, para. 45–46.
[26] *General Comment No. 19*, para. 47–51.
[27] *General Comment No. 14*, para. 5.

community are unlikely to be able to meet all health care requirements listed in *General Comment No. 14*. The same applies to the right to social security: Only a few industrialized countries have so far succeeded in offering their residents comprehensive social protection at an appropriate level.

However, the authors of the ICESCR have been well aware of this problem. Like any other social human right, therefore, the rights to health and social protection are limited by the "progression clause" in Art. 2 (1) ICESCR. According to this provision, states are obliged:

> to take steps, individually and through international assistance and co-operation, especially economic and technical, to the maximum of its available resources, with a view to achieving progressively the full realization of the rights recognized in the present Covenant by all appropriate means, including particularly the adoption of legislative measures.

This clause, which is important for the understanding of social rights,[28] takes account of the fact that in particular the governments of low-income countries often do not have the financial resources needed to implement their social rights obligations immediately. The Covenant thus differentiates between those states, which, due to their economic strength, can guarantee their citizens a level of protection equivalent to the requirements of the respective social rights, and those which are yet unable to do so.

The obligation of "progressive realization", in turn, is limited by the concept of "core obligations" which each State Party has to comply with—regardless of its economic resources—and which are specified in the respective *General Comments* for the social rights listed in the ICESCR. The core content of the right to health includes, among others, non-discriminatory access to health care, adequate basic food and drinking water supplies, adequate housing, access to key medicines, equitable distribution of all healthcare facilities, goods and services, and the development and implementation of a national health strategy.[29] Likewise, each state has to fulfill certain minimum requirements regarding the right to social security: social protection schemes must be made available to the whole population providing a minimum level of benefits to all persons that will enable them "to acquire at least

[28]For the interpretation of this clause see also CESCR, *General Comment no. 3 on the nature of States parties obligations (Art. 2, par.1), E/1991/23*, para. 9; Alston and Quinn (1987), pp. 172 et seq.; Sepúlveda (2003), pp. 174 et seq.; Young (2012), pp. 101 et seq.; Skogly (2012), advocating for a qualitative approach to the term 'resources' as provided in Article 2(1) of the ICESCR; Kendrick (2017), showing how the tools for measuring efficiency in a microeconomic sense can be applied to measure a duty-bearer's compliance with its obligation to fulfill social rights; cf. also Vandenhole (2016), p. 95, who critically notes that, in the case of the right to social security, the Committee assumes that the core obligations are also qualified by the 'maximum available resources'-clause.

[29]*General Comment No. 14*, para. 43; see also UN Committee on the Rights of the Child, *General Comment No. 15, The Right of the Child to the Enjoyment of the Highest Attainable Standard of Health (Art. 24)*, UN Doc. RC/C/GC/15 (2013), para. 73; for an analysis of the core obligations regarding the right to health see Tobin (2012), pp. 238 et seq.; Forman (2015).

essential healthcare, basic shelter and housing, water and sanitation, foodstuffs, and the most basic forms of education."[30]

But even these basic requirements are not met by many states—as the ILO and WHO figures mentioned above have clearly shown. At this point, the auxiliary obligation of the international community sets in: The so-called "extraterritorial obligation", which also follows from Art. 2 (1) ICESCR, requires that the state parties must also engage outside of their territory to implement the Covenant provisions. As far as their financial resources allow, richer members of the international community have to support poorer states in their efforts to implement, among others, the rights to health and social security.[31] This commitment to international assistance has also been explicitly reaffirmed in the *Maastricht Principles on Extraterritorial Obligations of States in the area of Economic, Social and Cultural Rights*.[32] The *Maastricht Principles* have been published in 2012 by the so-called ETO-Consortium (Extraterritorial Obligations-Consortium), a network of more than 140 human rights researchers and non-governmental organizations. Although the ETO-Consortium is only a civil society initiative, its influence on the interpretation of human rights obligations is by no means low, as reflected, for example, in the repeated reference to the *Maastricht Principles* in the recently published *General Comment No. 24*.[33] It is, however, still not specified to exactly what extent and in which relation between the potential donor states these extraterritorial obligations exist—neither in the *General Comments No. 14* and *19* nor in the *Maastricht Principles.* So far, international law provides only a general obligation to support poorer countries, the details are left to political negotiations.[34]

[30]*General Comment No. 19*, para. 59a.

[31]*General Comment No. 19*, para. 55 (see also para. 52, 58); *General Comment No. 14*, para. 39 (see also para. 45); Tobin (2012), p. 328.

[32]Art. 33 *Maastricht Principles*, http://www.etoconsortium.org/en/main-navigation/library/maastricht-principles/ (last accessed 27 March 2019); see De Schutter et al. (2012) and Langford et al. (2013).

[33]CESCR, *General comment No. 24 (2017) on State obligations under the International Covenant on Economic, Social and Cultural Rights in the context of business activities*, UN Doc. E/C.12/GC/24 (the Maastricht Principles are mentioned in footnote 71 and 78).

[34]Cf. on this Khalfan (2013), pp. 324 et seq.; Vandenhole and Benedek (2013), pp. 340 et seq.; Ssenyonjo (2016), pp. 118 et seq.; see also Vandenhole, in this volume.

3 International Political Initiatives to Improve the Implementation of the Right to Social Security and the Right to Health

The ILO Social-Protection Floor-Recommendation and the Global Partnership for Universal Social Protection to Achieve the Sustainable Development Goals (USP2030)

In 2012, the *International Labour Conference* adopted the *Social Protection Floor Recommendation*,[35] marking a new phase in the history of global social policy. If one compares the recommendation with *General Comment No. 19*, it becomes clear that ILO law (including ILO-soft law) and human rights in the field of social protection are now closely linked.[36] Both documents emphasize the key elements of the so-called "rights-based approach", which must be taken into account when setting up social protection systems.[37] In addition to providing a firm basis for social benefits in the national legal system, the approach includes a special focus on adequate participation of the population at all stages of the implementation process (legislation, program concretization and individual decisions by the authorities) and on monitoring and grievance mechanisms which enable civil society actors and especially persons directly affected to detect implementation deficiencies and, if necessary, defend themselves against them.

These procedural measures, however, will not have any effect for those people being not covered by the social protection systems. The ILO has also recognized this and therefore launched a new initiative to expand global social security at the turn of the millennium.[38] Until then, in its standard-setting activities—not least due to its

[35]*Recommendation concerning National Floors of Social Protection, ILO-Recommendation 102*; cf. Cichon et al. (2011). The ILO Committee of Experts on the Application of Conventions and Recommendations has recently published a report which clarifies the challenges that are associated with the implementation of the recommendation; see ILO (2019).

[36]Philip Alston, Special Rapporteur on extreme poverty and human rights, clearly emphasized this connection: "Implementation of the right to social protection through the adoption by all States of social protection floors is by far the most promising human rights-inspired approach to the global elimination of extreme poverty. In essence, those floors are guarantees of basic income security and access to essential social services for the whole population. No other operational concept has anything like the same potential to ensure that the poorest 15 to 20% of the world's people enjoy at least minimum levels of economic, social and cultural rights."; see *Report of the Special Rapporteur on extreme poverty and human rights to the UN General Assembly* (of 11 August 2014), UN Doc. A/69/297, para. 2, https://www.ohchr.org/EN/Issues/Poverty/Pages/AnnualReports.aspx (last accessed 27 March 2019).

[37]Cf. Kaltenborn (2017b), pp. 250 et seq.; on the rights-based approach to social protection see in detail Sepúlveda and Nyst (2012); for a general introduction to the human rights-based approach to development see Fukuda-Parr (2016), p. 203.

[38]Deacon (2013), pp. 28 et seq.

tripartite structure[39]—the organization had been mainly concerned with the social protection of those people who are in formal employment. However, a large part of the working-age population of the Global South works in the informal sector and is therefore not covered by traditional security systems.[40] The main purpose of the Social Protection Floor-initiative is therefore to provide states with a guideline for eliminating these and other gaps in the implementation of the right to social security.

The recommendation, adopted by consensus,[41] proposes that ILO member states introduce a social protection floor that guarantees access to basic health care and basic income support for all residents; in addition, the states are called upon to continuously raise this basic protection to the level of already existing ILO standards.[42] From a human rights perspective, the obligation to provide for a social protection floor can thus be considered largely equivalent to the core obligations which the ICESCR state parties must observe regarding the right to social security (see Sect. 2). It remains up to their national social policy strategy which instruments governments use to achieve this basic protection—whether they set up social insurance schemes (and, where appropriate, include also private providers) or mainly rely on social assistance programs.[43] The level of social benefits must be regularly reviewed in a legally defined transparent procedure.[44] Social protection programs should primarily be financed from domestic funds; if a country is unable to do so for economic reasons, it may seek temporary international support.[45] The *Social Protection Floor Recommendation* has been endorsed by many other international actors, including G20,[46] the European Union,[47] several other UN institutions[48] and

[39]The decision-making bodies of the ILO are composed of government, employee and employer representatives.

[40]See generally World Bank (2019), pp. 94 et seq.; ILO (2017b), pp. 173 et seq.; UN DESA (2018), p. 19; Alfers et al. (2017); Rutkowski (2018); cf. in this context also the work of the International Social Security Association (2016) which has formulated administrative solutions to improve access to contributory social security programs for populations that are difficult to cover.

[41]http://www.ilo.org/brussels/WCMS_183640/lang%2D%2Den/index.htm (last accessed 27 March 2019).

[42]*ILO-Recommendation 102*, para. 13.

[43]*ILO-Recommendation 102*, para. 9.

[44]*ILO-Recommendation 102*, para. 7 c).

[45]*ILO-Recommendation 102*, para 12.

[46]*G20 Cannes Summit Final Declaration* (of 4 November 2011), para. 4, 77, http://www.mofa.go.jp/policy/economy/g20_summit/2011/declaration.html (last accessed 27 March 2019); see also *G20 2013 St. Petersburg Summit Leaders' Declaration* (of 9 November 2013), para. 30 https://www.g20germany.de/Content/DE/_Anlagen/G7_G20/G20-erklaerung-petersburg-en___blob=publicationFile&v=1.pdf (last accessed 27 March 2019).

[47]*Communication of the EU-Commission on Social Protection in European Union Development Cooperation* (of 20 August 2012), COM(2012) 446 final, S. 6.

[48]WHO (2010), p. 11; UNDP and ILO (2011); UNICEF (2012), p. 32; World Bank (2012), p. 14; HLPE (2012), p. 25; FAO (2017), p. 19.

also civil society,[49] and thus, even though it is only a soft law instrument, it can be regarded—in addition to the ICESCR—as the main international legal document for the global implementation of the right to social security.

In 2016 the World Bank and ILO decided to launch a joint initiative aiming to further develop the idea of the Social Protection Floor and to provide social protection for the entire population of a country. The *Global Partnership for Universal Social Protection to Achieve the Sustainable Development Goals*[50] (since 2017 it has the name "USP2030") provides a platform for joint action on the establishment of comprehensive, integrated social protection systems. At UN level, the USP2030-partnership also involves UNICEF, the United Nations Development Programme (UNDP) and the Food and Agriculture Organization of the United Nations (FAO), other partners are e.g. the African Union, the European Commission, OECD, Helpage International, the International Council on Social Welfare (ICSW), the International Social Security Association (ISSA) and several development agencies.

The International Health Partnership for UHC 2030

In global health policy, the issue of Universal Health Coverage (UHC) has significantly shaped the debates in recent years.[51] Already in 2005 the WHO invited its Member States

> to ensure that health-financing systems include a method for prepayment of financial contributions for health care, with a view to sharing risk among the population and avoiding catastrophic health-care expenditure and impoverishment of individuals as a result of seeking care; to ensure adequate and equitable distribution of good-quality health care infrastructures and human resources for health so that the insurees will receive equitable and good-quality health services according to the benefits package; . . . to plan the transition to universal coverage of their citizens so as to contribute to meeting the needs of the population for health care and improving its quality, to reducing poverty, . . . and to achieving health for all; . . .[52]

Since then, a number of other resolutions have been adopted at the international level specifically addressing UHC. The topic not only concerns the United

[49]For the work of the Global Coalition for Social Protection Floors see http://www.socialprotectionfloorscoalition.org/ (last accessed 27 March 2019).

[50]https://www.usp2030.org/gimi/USP2030.action; see also World Bank Group and ILO (2016); BMZ (2019).

[51]Gentilini (2018) rightly points out that there are many similarities in terms of strategies to achieve universal coverage with regard to the health sector and to social protection.

[52]World Health Assembly, *Sustainable health financing, universal coverage and social health insurance* (of 25 May 2005), WHA58/2005/REC/1; see also World Health Assembly, *Sustainable health financing structures and universal coverage* (of 24 May 2011), WHA64.9/2011/REC/.

Nations,[53] but also increasingly the G7- and G20-summits.[54] According to the definition given by the WHO

> "UHC means that all individuals and communities receive the health services they need without suffering financial hardship. It includes the full spectrum of essential, quality health services, from health promotion to prevention, treatment, rehabilitation, and palliative care. UHC enables everyone to access the services that address the most important causes of disease and death, and ensures that the quality of those services is good enough to improve the health of the people who receive them. ... UHC is not only about ensuring a minimum package of health services, but also about ensuring a progressive expansion of coverage of health services and financial protection as more resources become available. ..."; UHC does however "not mean free coverage for all possible health interventions, regardless of the cost, as no country can provide all services free of charge on a sustainable basis".[55]

Here, too, there are partial overlaps with the core obligations which must be observed regarding the right to health (see Sect. 2).[56]

Already in 2007, several stakeholders joined forces to coordinate efforts to develop national health systems as part of a global partnership and to advance the expansion of health protection in the Global South. Among the 26 signatories of the *International Health Partnership (IHP+) Global Compact* were seven states in which the UHC objectives are to be implemented,[57] furthermore bilateral and multilateral donor organizations[58] as well as global Public Private Partnerships[59] and private foundations.[60] Meanwhile, the UHC alliance has grown to 118 participants[61]; additional members (especially middle-income countries and private sector representatives) are recruited. Since 2016, it also has a new name: In order to clarify

---

[53]UN General Assembly, *Global health and foreign policy* (of 6 December 2012), UN Doc. A/RES/ 67/81.

[54]*G7 Ise-Shima Vision for Global Health* (of 27 May 2016), para. 2-1-2, http://www.mofa.go.jp/files/000160273.pdf (last accessed 27 March 2019); *Berlin Declaration of the G20 Health Ministers "Together Today for a Healthy Tomorrow"* (of 20 May 2017), para. 17 et seq., http://www.bundesgesundheitsministerium.de/fileadmin/Dateien/3_Downloads/G/G20-Gesundheitsministertreffen/G20_Health_Ministers_Declaration_engl.pdf (last accessed 27 March 2019).

[55]http://www.who.int/mediacentre/factsheets/fs395/en/ (last accessed 27 March 2019). On the multi-dimensional concept of UHC (legal, humanitarian social, health economics, and public health concept) see Abiiro and De Allegri (2015). One of the main questions when introducing UHC is to decide which services are critical to a successful health benefits package; see Glassman et al. (2017).

[56]*Report of the Special Rapporteur on the right of everyone to the enjoyment of the highest attainable standard of physical and mental health* (of 5 August 2016), UN Doc. A/71/304, para. 28, 74–90; for a detailed comparison of the UHC-concept with the right to health see also Ooms et al. (2014); Forman et al. (2016); Chapman (2016), pp. 283–326.

[57]Burundi, Cambodia, Ethiopia, Kenya, Mozambique, Nepal, Zambia.

[58]Canada, France, Germany, Italy, Netherlands, Norway, Portugal, UK, European Commission (EC), African Development Bank (AfDB), UNAIDS, UNDP, UNFPA, UNICEF, WHO, World Bank.

[59]GAVI Vaccine Alliance (GAVI), Global Fund to Fight AIDS, TB and Malaria (Global Fund).

[60]The Bill & Melinda Gates Foundation.

[61]https://www.uhc2030.org/about-us/uhc2030-partners/ (last accessed 27 March 2019).

the transition to the new development paradigm of the Sustainable Development Goals (SDGs), it is now called *International Health Partnership for UHC 2030 (UHC2030)*.[62] The core goal of the UHC alliance continues to be the effectiveness of health-related development cooperation (based on the *Paris Declaration on Aid Effectiveness*[63]). In addition, however, the effective use of domestic resources has increasingly become the focus of the work of the initiative. The UHC alliance itself does not act as a funding organization, but merely as a "platform for discussions on global health and health system strengthening and as a place for mutual accountability drawing on existing initiatives".[64] The main instruments for the implementation of the UHC-goals are "country compacts which work together more effectively to improve aid effectiveness and deliver priorities in the national health strategy"[65]; furthermore, so-called Joint Assessments of National Health Strategies (JANS) are being undertaken to improve the quality and effectiveness of health strategies at the country level.[66]

Proposal for a Framework Convention on Global Health

Ensuring universal health care delivery is also one of the goals that the initiators of a new legislative project on Global Health Governance seek to achieve. The idea of a *Framework Convention on Global Health* (FCGH)[67] which has been primarily promoted by Lawrence O. Gostin and Eric A. Friedman of the O'Neill Institute of Georgetown University (Washington DC) has the aim of better coordinating international activities in the health sector via a legally binding instrument under international law, of specifying the obligations of states and other transnational actors to implement health-related development goals (in particular SDG 3), and of strengthening the human rights basis of global health law. Key elements of such a framework convention include binding (and verifiable) standards for the implementation and funding of health goals as well as the involvement of civil society in all relevant decision-making processes. Moreover, the extraterritorial human rights obligations

[62]The draft text of the updated *Global Compact* can be downloaded at https://www.uhc2030.org/fileadmin/uploads/ihp/Documents/About_IHP_/mgt_arrangemts___docs/UHC_Alliance/Official_documents_2017/UHC2030_TSC_Global_Compact_Jan_2017_WIP.pdf (last accessed 27 March 2019). For an assessment of IHP+ see Shorten et al. (2012).

[63]http://www.oecd.org/dac/effectiveness/34428351.pdf (last accessed 27 March 2019).

[64]https://www.uhc2030.org/fileadmin/uploads/ihp/Documents/About_IHP_/mgt_arrangemts___docs/Steering_Committee_as_of_2014/SC_VI/IHP__Note_for_the_Record_Sixth_SC_meeting_April_2016.pdf (last accessed 27 March 2019).

[65]https://www.uhc2030.org/what-we-do/coordination-of-health-system-strengthening/country-compact/ (last accessed 27 March 2019).

[66]https://www.uhc2030.org/what-we-do/coordination-of-health-system-strengthening/jans-tool-and-guidelines/ (last accessed 27 March 2019).

[67]https://fcghalliance.org/ (last accessed 27 March 2019).

of states and the links to other policy areas (such as trade, protection of intellectual property) should be addressed in the FCGH.[68]

Undoubtedly, a carefully elaborated framework convention containing, in particular, precisely formulated legal requirements[69] would be a great advantage, not only from the perspective of global health governance. It could also make an important contribution to the debate on the constitutionalization of international law: For global health law—a cross section subject covering several subareas of international law—the standard-setting role of relevant human rights (in this case the rights to health and social protection) could be made much more explicit than it has previously been the case. The project is ambitious—but perhaps too ambitious to have serious chances of being realized in the foreseeable future. It seems rather doubtful whether within the next years a consensus could emerge among governments to move closer to this proposal. So far, neither the G7 group nor any other stakeholders have shown any willingness to establish new international hard law in the field of health financing, exceeding the existing soft law obligations.

4 Social Protection Floors and Universal Health Coverage in the 2030 Sustainable Development Agenda: Financial Responsibilities of the International Community

Since 2015, the global calls for establishing social protection floors and extending health coverage have become an integral part of the *2030 Agenda for Sustainable Development*.[70] Under the heading "End poverty in all its forms everywhere", reference is expressly made to the terminology of the *SPF Recommendation*. According to SDG 1.3 states have agreed to

> (i)mplement nationally appropriate social protection systems and measures for all, including floors, and by 2030 achieve substantial coverage of the poor and the vulnerable.[71]

---

[68]Cf. Friedman and Gostin (2012), p. 4; Buse et al. (2014); Gostin (2014), p. 437 ff.; Friedman (2016); for a critical analysis of the FCGH project see Hoffmann and Rottingen (2013).

[69]See for the advantages of using the framework agreement-model Toebes (2015), p. 19 et seq.

[70]*Transforming our world: The 2030 Agenda for Sustainable Development*, UN Doc. A/RES/70/1, https://sustainabledevelopment.un.org/post2015/transformingourworld (last accessed 27 March 2019).

[71]The indicator for social protection coverage (indicator 1.3.1) is the proportion of population covered by social protection floors/systems, disaggregated by sex, and distinguishing children, unemployed, old age, people with disabilities, pregnant women/new-borns, work injury victims, poor and vulnerable.—The importance of SDG 1.3 has recently been reiterated by the Commission for Social Development (of 20 November 2018), UN Doc. E/CN.5/2019/3.

A group of high-level UN human rights experts had even campaigned to address social protection as an independent new development goal in the new Agenda.[72] Although this was not successful in the end, it is clear that social protection is now one of the key issues of the new catalogue of global goals. It has a kind of "bridging function" between various goals because social protection programs are not only an important tool in fighting poverty (SDG 1.1, 1.2, 1.5), but also the basis for appropriate health care and food security (SDG 1.5, 2.1, 2.2, 3.4, 3.8), for social cohesion (SDG 10.2), for the reduction of inequality (SDG 4.5, 5.1, 5.4, 10.1, 10.4) and for helping people back into work (SDG 8.5, 8.6). Moreover, social protection is a prerequisite for enabling parents to send their children to school instead of encouraging them to contribute to the household income (SDG 8.7).[73]

Universal Health Coverage is also given high priority in the *2030 Agenda*. While some significant progress in global health care has been already made in the period between 2000 and 2015, especially in those sectors explicitly identified as Millennium Development Goals (MDG 4: "reducing child mortality", MDG 5: "improving maternal health", and MDG 6: "combating HIV/AIDS, malaria and other diseases"),[74] there are still severe deficiencies in general health care in many developing and newly industrializing countries. The *2030 Agenda* therefore once more takes up health policy issues—now listed in SDG 3—and calls on states, among other things:

> to "(a)chieve universal health coverage, including financial risk protection, access to quality essential health-care services and access to safe, effective, quality and affordable essential medicines and vaccines for all" (SDG 3.8).

The heading of SDG 3 explicitly clarifies that this ambitious goal should be achieved "for all at all ages".[75] SDG 3.8 is therefore closely related to SDG 1.3, which also includes access for all people to essential health care services as part of the nationwide expansion of social protection programs.

As with most social rights, the right to health and the right to social security, too, can only be achieved by the provision of substantial financial and technical resources. Similar to the *Social Protection Floor Recommendation*, the *2030 Agenda* emphasizes that first and foremost each country for itself has the primary

---

[72]Press Release of the Office of the United Nations High Commissioner for Human Rights (of 21 May 2013), http://newsarchive.ohchr.org/EN/NewsEvents/Pages/DisplayNews.aspx?NewsID=13342&LangID=E (last accessed 27 March 2019).

[73]Kaltenborn (2015), p. 3; BMZ (2017), pp. 4, 11; see also Loewe and Strupat (2017).

[74]See *United Nations Millennium Declaration* (of 18.9.2000), UN Doc. A/RES/55/2.

[75]The Special Rapporteur on the right to health critically points out that target 3.8 does "not make explicit commitments to confer priority to the poor and marginalized either in the process of expanding coverage or in developing priorities as to which services to provide. Without those clear commitments, there is a risk that universal health coverage efforts will entrench inequality" and, as a consequence, could be inconsistent with human rights requirements; see *Report of the Special Rapporteur on the right of everyone to the enjoyment of the highest attainable standard of physical and mental health* (of 5 August 2016), UN Doc. A/71/304, para. 76; see also Puras (2016), p. 8.

responsibility for achieving the development goals, including SDGs 1.3 and 3.8.[76] It is the task of the respective government and parliament to implement these goals—and, by this means, also the corresponding human rights obligations—through appropriate legislative acts as well as universal health and social policy programs. In the recent past, a number of countries have shown that they are capable of independently developing their social and health systems to meet basic human rights requirements. Notable advances have been made recently, particularly in some countries which rank as middle-income countries according to the World Bank classification. For instance, Brazil and China have launched ambitious reform programs such as the introduction of cash-transfer schemes[77] or the nationwide expansion of pensions[78]—by this way, both countries played a decisive part in reaching MDG 1 (halving extreme poverty) five years earlier before expiration of the MDG deadline.[79] But in recent years, a range of poorer members of the international community, too, have been successful in combating poverty due to their newly established (partly donor-financed) cash transfer programs.[80] Moreover, some countries have already successfully established nationwide basic health care programs, like, among others, China, Colombia, Rwanda and Thailand.[81]

However, the figures on global social security and health care gaps mentioned above have made it obvious that a large number of states have difficulties in making the necessary funds available from their own financial resources to achieve SDGs 1.3 and 3.8.[82] The *2030 Agenda* therefore emphasizes that

> international public finance plays an important role in complementing the efforts of countries to mobilize public resources domestically, especially in the poorest and most vulnerable countries with limited domestic resources. An important use of international public finance, including official development assistance (ODA), is to catalyse additional resource mobilization from other sources, public and private. ODA providers reaffirm their respective commitments, including the commitment by many developed countries to achieve the target

[76] *2030-Agenda*, para. 41.

[77] The best-known of the Brazilian programs in the area of basic social protection is the program Bolsa Família, see Robles and Mirosevic (2013).

[78] World Bank/ILO (2018), p. 53.

[79] http://www.un.org/millenniumgoals/poverty.shtml (last accessed 27 March 2019).

[80] Cf., inter alia, Garcia and Moore (2012); Barrientos (2013); Bastagli et al. (2016).

[81] ILO (2017b), p. 116; cf. also the comparative study on UHC programs in 24 countries presented by the World Bank (2015); for recent data on the two UHC-indicators (3.8.1: Coverage of essential health services, defined as the average coverage of essential services based on tracer interventions that include reproductive, maternal, newborn and child health, infectious diseases, non-communicable diseases and service capacity and access, among the general and the most disadvantaged population; and 3.8.2: Proportion of population with large household expenditures on health as a share of total household expenditure or income), see WHO and World Bank (2017); for an analysis of the special challenges which African countries face regarding UHC, see Dovlo, in this volume.

[82] However, it is partly the lack of willingness on the part of governments to set corresponding priorities in national budgetary policy. On re-prioritizing strategies aiming at expanding fiscal space for social protection programs see Ortiz et al. (2017), p. 5.

of 0.7 per cent of gross national income for official development assistance (ODA/GNI) to developing countries and 0.15 per cent to 0.2 per cent of ODA/GNI to least developed countries.[83]

The development of health systems has been supported with financial and technical means of donor organizations and partner countries already for a long time.[84] One reason for the fact that the health sector's proportion of the total budget of development cooperation turns out to be this large might be the strong involvement of some of the largest philanthropic foundations that have invested significantly more than half of their total budget in this sector in recent years.[85] But the governments of several industrialized countries also provide considerable funds for the health sector in developing countries.[86] Just to take an example: the German government provided € 473 million in bilateral development assistance to the health sector in 2017, furthermore € 19.9 million have been directed to the WHO, € 60 million to GAVI and € 230 million to the Global Fund.[87] But it must not be overlooked that these financial commitments are sometimes at least partially guided by self-interest. The outbreak of epidemics and pandemics (e.g. Ebola, Zika) in developing countries also poses a threat to people living in industrialized countries. Therefore, building effective health systems in the South is also in the interest of other countries as it facilitates effective cross-border health protection.[88]

The development and expansion of those parts of social protection systems in the Global South which are not directly health-related is also supported by the donor community with financial and technical means, but to a much lesser extent than the health sector.[89] On the one hand, this is certainly due to the fact that social protection

---

[83]*2030-Agenda*, para. 43.

[84]For an analysis of the problems currently under discussion see Moon and Omole (2017).

[85]OECD (2017), p. 123. In 2015 the Bill & Melinda Gates Foundation was the third-largest provider in the health and reproductive health sector (with USD 3.3 billion of disbursements; see ibid. p. 155).

[86]OECD (2017), p. 151, cf. also http://www.oecd.org/dac/financing-sustainable-development/development-finance-topics/aidtohealth.htm (last accessed 27 March 2019).

[87]http://www.bmz.de/de/ministerium/zahlen_fakten/oda/leistungen/bilaterale_oda_foerderbereiche_laender_2017/index.html; http://www.bmz.de/de/ministerium/zahlen_fakten/oda/leistungen/deutsche_ODA_EU_2013_2017/index.html (last accessed 17 November 2019).

[88]Cf. in this context the *G 7 Ministerial Declaration "Beyond Ebola: A G7 agenda to help prevent future crises and enhance security in Africa"* of 15 April 2015, p. 2, https://www.auswaertiges-amt.de/en/newsroom/news/150415-g7-beyond-ebola/270868 (last accessed 27 March 2019).

[89]According to a study from 2013, approximately $ 3.4 billion of ODA were invested in social protection programs (excluding food aid); see http://devinit.org/post/financing-social-protection-ldcs/ (last accessed 27 March 2019). In the same year, the health sector was funded with nearly $ 13 billion; see http://www.oecd-ilibrary.org/development/data/oecd-international-development-statistics/official-bilateral-commitments-by-sector_data-00073-en (last accessed 27 March 2019). However, it is difficult to provide reasonably accurate figures for such a comparison, since projects designed to strengthen social protection systems are generally not separately identifiable, but can be allocated to different other parts of the ODA; see http://www.oecd.org/dac/povertyreduction/Evolution%20of%20ODA%20for%20Social%20Protection.pdf (last accessed 27 March 2019).

is an area of development cooperation that has only recently received increased global attention. Another reason for the reluctance of the donor community presumably lies in the fact that social protection programs are, at least in part, investments whose effects are more likely to occur in the medium to long term (the most obvious example are pension insurances) and therefore appear less urgent to donors than other projects financed by development assistance. It should not be forgotten, however, that social protection programs play an important role not only in the fight against poverty but also in the pursuit of many other goals of the *2030 Agenda* (the "bridging function" of SDG 1.3 has already been mentioned). Especially against the background of some global political events of the recent past—on the one hand the various climate-related disasters, on the other hand the refugee crisis—the growing importance of long-term income security programs that cover the entire life cycle of people becomes clear: In countries that are particularly affected by drought, hurricanes or heavy storms due to climate change, social programs that protect people from the sudden loss of livelihoods become more important.[90] And ultimately, the issue is also important in the context of the debate on the refugee crisis: As long as the labour market does not provide sufficient opportunities for income security, people will have (albeit modest) reasons to remain in the country only if they can escape extreme poverty through basic social assistance programs and if they can be confident to be covered by adequate social protection also in old age.[91]

Therefore, the revitalization of the "Global Partnership" in SDG 17[92] is of great significance to global social protection—which means that further substantial efforts will be needed,[93] in addition to the development cooperation programs already existing in this sector. Without the support of foreign partners, many governments in the Global South still have great difficulties to finance basic social protection including essential health services for their citizens. Under international law, as has been shown above (see Sect. 2), there is even a legal (extraterritorial) obligation of the wealthier members of the international community to provide this support and thus contribute to the global implementation of the right to social security and to the right to health.

5 Conclusion

In its General Comments No. 14 and 19, the CESCR has precisely specified the contents of the right to health and the right to social security. The main challenges associated with the implementation of these two human rights have been addressed

---

[90]Cf. Hallegate et al. (2017), pp. 148 et seq.

[91]However, this topic still requires intensive further research; cf. on this e.g. Gesellschaft für Versicherungswissenschaft und –gestaltung (2017); Adhikari and Gentilini (2018).

[92]See also Martens and Vandenhole, in this volume.

[93]For example, it has been suggested that a Global Fund should be established which could help countries to set up or extend their social protection systems; cf. the references in Kaltenborn (2015), p. 3.

in several major international policy initiatives and global partnerships (the Social Protection Floors Initiative, the Universal Social Protection Partnership and the International Health Partnership for Universal Health Coverage). The 2030 Agenda now makes an important contribution to the concretization of the rights to health and social security, because it expressly obliges the international community both to implement the concept of social protection floors and to ensure universal health protection.

References

Abiiro AA, De Allegri M (2015) Universal health coverage from multiple perspectives: a synthesis of conceptual literature and global debates. BMC Int Health Hum Rights 15:17

Adhikari S, Gentilini U (2018) Should I stay or should I go. Do cash transfers affect migration? Policy Research Working Paper 8525, The World Bank, Washington, DC. http://documents.worldbank.org/curated/en/609571531402897490/pdf/WPS8525.pdf. Accessed 27 Mar 2019

Alfers L, Lund F, Moussié R (2017) Approaches to social protection for informal workers: aligning productivist and human rights-based approaches. Int Soc Secur Rev 70(4):67–85

Alston P, Quinn G (1987) The nature and scope of states parties' obligations under the International Covenant on Economic, Social and Cultural Rights. Hum Rights Q 9:156–229

Barrientos A (2013) Social assistance in developing countries. Cambridge University Press, Cambridge

Bastagli F, Hagen-Zanker J, Harman L, Barca V, Sturge G, Schmidt T (2016) Cash transfers. What does the evidence say? A rigorous review of programme impact and of the role of design and implementation features. Overseas Development Institute, London

Bender K (2016) Soziale Sicherung und internationale Politikdiffusion. MenschenRechtsMagazin Heft 2/2016, pp 129–138

Bender K, Keller S, Willing H (2014) The role of international policy transfer and diffusion for policy change in social protection – a review of the state of the art. BMZ/GIZ/IZNE, Bonn. https://www.h-brs.de/files/izne/policy_diffusion_1.pdf. Accessed 27 Mar 2019

Braveman P (2010) Social conditions, health equity, and human rights. Health Hum Rights J 12 (2):31–48

Bundesministerium für wirtschaftliche Zusammenarbeit und Entwicklung (BMZ) (2017) Social protection for equitable development. BMZ, Bonn. http://www.bmz.de/en/publications/type_of_publication/strategies/Strategiepapier420_09_2017.pdf. Accessed 27 Mar 2019

Bundesministerium für wirtschaftliche Zusammenarbeit und Entwicklung (BMZ) (2019) A systems perspective on Universal Social Protection. Towards life-long equitable access to comprehensive social protection for all. BMZ, Bonn. http://health.bmz.de/ghpc/discussion_papers/Universal_Social_Protection/GHPC_USP_Final-1.pdf. Accessed 27 Mar 2019

Buse K, Gostin LO, Friedman EA (2014) Pathways towards a Framework Convention on Global Health. In: Freeman M, Hawkes S, Bennett B (eds) Law and global health. Oxford University Press, Oxford, pp 37–62

Chapman AR (2010) The social determinants of health, health equity, and human rights. Health Hum Rights J 12(2):17–30

Chapman AR (2016) Global health, human rights and the challenge of neoliberal policies. Cambridge University Press, Cambridge

Cichon M, Behrendt C, Wodsak V (2011) The UN social protection floor initiative: moving forward with the extension of social security. Internationale Politik und Gesellschaft (IPG), 2012/2, pp 32–50

Coomans F (2008) Justiciability of economic and social rights. Experiences from domestic systems. Intersentia, Antwerp
Damrosch LF (2011) Gross and systematic human rights violations. Max planck encyclopedia of public international law. Oxford University Press, Oxford
De Schutter O, Eide A, Khalfan A, Orellana M, Salomon M, Seiderman I (2012) Commentary to the Maastricht principles on extraterritorial obligations of states in the area of economic, social and cultural rights. Hum Rights Q 34:1084–1169
Deacon B (2013) Global social policy in the making. Policy Press, Bristol
Ferguson L (2017) Assessing work at the intersection of health and human rights: why, how and who? In: Andreassen BA, Sano H-O, McInerney-Lankford S (eds) Research methods in human rights. A handbook. E. Elgar, Cheltenham, pp 408–434
Food and Agriculture Organization of the United Nations (FAO) (2017) FAO social protection framework. FAO, Rome
Forman L (2015) Can core obligations under the right to health achieve their ambitions? Zeitschrift für Menschenrechte (J Hum Rights) 9(2):36–48
Forman L, Beiersmann C, Brolan CE et al (2016) What do core obligations under the right to health bring to universal health coverage? Health Hum Rights J 18(2):23–34
Forsythe DP (2018) Human rights in international relations, 4th edn. Cambridge University Press, Cambridge
Friedman EA (2016) An independent review and accountability mechanism for the Sustainable Development Goals: the possibilities of a framework convention on global health. Health Hum Rights J 18(1):129–140
Friedman EA, Gostin LO (2012) Pillars for progress on the right to health: harnessing the potential of human rights through a Framework Convention on Global Health. Health Hum Rights J 14 (1):1–16
Fukuda-Parr S (2016) Human rights and politics in development. In: Goodhart M (ed) Human rights: politics and practice, 3rd edn. Oxford University Press, Oxford, pp 198–215
Garcia M, Moore CMT (2012) The cash dividend – the rise of cash transfer programs in Sub-Saharan Africa. The World Bank, Washington DC
Gentilini U (2018) What lessons for social protection from universal health coverage? Let's Talk Development blog, World Bank, August 22. http://blogs.worldbank.org/developmenttalk/what-lessons-social-protection-universal-health-coverage. Accessed 27 Mar 2019
Gesellschaft für Versicherungswissenschaft und -gestaltung e.V. (2017) Social protection as an alternative to migration? An assessment of the role of social protection in reducing push factors for migration in different country contexts. Executive summary. GVG, Berlin
Glassman A, Giedion U, Smith PC (2017) What's in, what's out: designing benefits for universal health coverage. Brookings Institution Press, Washington DC
Gloppen S (2008) Litigation as a strategy to hold governments accountable for implementing the right to health. Health Hum Rights J 10(2):21–36
Goldblatt B (2016) Developing the right to social security - a gender perspective. Routledge, Abingdon/New York
Gostin LO (2014) Global health law. Harvard University Press, Cambridge/London
Hallegate S, Vogt-Schilb A, Bangalore M, Rozenberg J (2017) Unbreakable – building the resilience of the poor in the face of natural disasters. The World Bank, Washington DC
High Level Panel of Experts on Food Security and Nutrition (HLPE) (2012) Social protection for food security. Committee on World Food Security, Rome. http://www.fao.org/fileadmin/user_upload/hlpe/hlpe_documents/HLPE_Reports/HLPE-Report-4-Social_protection_for_food_security-June_2012.pdf. Accessed 27 Mar 2019
Hoffmann SJ, Rottingen JA (2013) Dark sides of the proposed Framework Convention on Global Health's many virtues: a systematic review and critical analysis. Health Hum Rights J 15 (1):117–134
Human Rights Watch (2018) World Report 2018, Events of 2017. https://www.hrw.org/sites/default/files/world_report_download/201801world_report_web.pdf. Accessed 27 Mar 2019

International Labour Office (ILO) (2017a) Building social protection systems. International standards and human rights instruments. ILO, Geneva
International Labour Office (ILO) (2017b) World Social Protection Report 2017-19. Universal social protection to achieve the Sustainable Development Goals. ILO, Geneva
International Labour Office (ILO) (2019) General survey concerning the Social Protection Floors Recommendation, 2012 (No. 202), ILC.108/III/B, ILO, Geneva
International Social Security Association (ISSA) (2016) Administrative solutions for extending coverage. ISSA, Geneva
Kaasch A (2015) Shaping global health policy. Global social policy actors and ideas about health care system. Palgrave Macmillan, Basingstoke (Hampshire)
Kaltenborn M (2015) Global social protection. New impetus from the 2030 Agenda for Sustainable Development. In: Stiftung Entwicklung und Frieden (ed) Global Governance Spotlight 7/2015. https://www.sef-bonn.org/fileadmin/SEF-Dateiliste/04_Publikationen/GG-Spotlight/2015/ggs_2015-07_en.pdf. Accessed 27 Mar 2019
Kaltenborn M (2017a) Das Recht auf soziale Sicherheit – Neue völkerrechtliche Entwicklungen und globale Herausforderungen. Zeitschrift für Menschenrechte (J Hum Rights) 11(2):88–101
Kaltenborn M (2017b) Overcoming extreme poverty by social protection floors – approaches to closing the right to social security gap. Law Dev Rev 10(2):237–273
Kaltenborn M, Troeppner D (2017) Globales Gesundheitsrecht. JuristenZeitung (JZ), pp 745–754
Kendrick A (2017) Measuring compliance: social rights and the maximum available resources dilemma. Hum Rights Q 39:657–679
Khalfan A (2013) Division of responsibility between states. In: Langford M, Vandenhole W, Scheinin M, van Genugten W (eds) Global justice, state duties. The extraterritorial scope of economic, social, and cultural rights in international law. Cambridge University Press, Cambridge, pp 291–331
Koehler G (2017) The politics of rights-based, transformative social policy in South and Southeast Asia. Int Soc Secur Rev 70(4):105–126
Krennerich M (2015) Das Menschenrecht auf Gesundheit. Zeitschrift für Menschenrechte (J Hum Rights) 9(2):8–35
Kuhlmann E, Blank R, Bourgeault I, Wendt C (2015) The Palgrave international handbook of healthcare policy and governance. Palgrave Macmillan, Basingstoke (Hampshire)
Lagoutte S, Gammeltoft-Hansen T, Cerone J (eds) (2016) Tracing the roles of soft law in human rights. Oxford University Press, Oxford
Langford M (2008) Social rights jurisprudence. Emerging trends in international and comparative law. Cambridge University Press, Cambridge
Langford M, Vandenhole W, Scheinin M, van Genugten W (2013) Global justice, state duties. The extraterritorial scope of economic, social, and cultural rights in international law. Cambridge University Press, Cambridge
Loewe M, Strupat (2017) No stability without social protection. The Current Column of 23 October 2017. German Development Institute/Deutsches Institut für Entwicklungspolitik (DIE), Bonn. https://www.die-gdi.de/uploads/media/German_Development_Institute_Loewe_Strupat_27.10.2017.pdf. Accessed 27 Mar 2019
Moon S, Omole O (2017) Development assistance for health. Critiques, proposals and prospects for change. Health Econ Policy Law 12:207–221
OECD (2017) Development Co-operation Report 2017: data for development. OECD Publishing, Paris
Ooms G, Latif LA, Waris A, Brolan CE, Hammonds R, Friedman EA, Mulumba M, Forman L (2014) Is universal health coverage the practical expression of the right to health care? BMC Int Health Hum Rights 14:3
Ortiz I, Cummins M, Karunanethy K (2017) Fiscal space for social protection and the SDGs: options to expand social investments in 187 countries. ESS Working Paper No. 48. ILO, Geneva
Puras D (2016) Universal health coverage: a return to Alma-Ata and Ottawa. Health Hum Rights J 18(2):7–10

Robles C, Mirosevic V (2013) Social protection systems in Latin America and the Caribbean: Brazil. Economic Commission for Latin America and the Caribbean, Santiago (Chile)
Rodgers G, Lee E, Swepston L, van Daele J (2009) The International Labour Organization and the quest for social justice, 1919–2009. ILO, Geneva
Rutkowski M (2018) Reimagining social protection. New systems are needed that do not rely on standard employment contracts. Finance Dev:10–13
Sepúlveda M (2003) The nature of the obligations under the International Covenant on Economic, Social and Cultural Rights. Intersentia, Antwerp
Sepúlveda M, Nyst C (2012) The human rights approach to social protection. Ministry for Foreign Affairs of Finland, Helsinki
Shelton DL (2014) Advanced introduction to international human rights law. E. Elgar, Cheltenham
Shorten T, Taylor M, Spicer N, Mounier-Jack S, McCoy D (2012) The International Health Partnership Plus: rhetoric or real change? Results of a self-reported survey in the context of the 4th high level forum on aid effectiveness in Busan. Glob Health 8:13
Skogly S (2012) The requirement of using the 'maximum of available resources' for human rights realisation: a question of quality as well as quantity? Hum Rights Law Rev 12(3):393–420
Ssenyonjo M (2016) Economic, social and cultural rights in international law, 2nd edn. Hart, Oxford
Tobin J (2012) The right to health in international law. Oxford University Press, Oxford
Toebes B (2015) The framework convention on global health: considerations in light of international law. Glob Health Gov IX(1):7–23
United Nations Department of Economic and Social Affairs (UN DESA) (2018) Promoting inclusion through social protection. Report on the World Social Situation 2018, UN DESA, New York. https://www.un.org/development/desa/dspd/wp-content/uploads/sites/22/2018/07/1-1.pdf. Accessed 27 Mar 2019
United Nations Development Programme (UNDP), International Labour Office (ILO) (2011) Sharing innovative experiences: successful social protection floor experiences, vol 18. UNDP, New York. http://www.ilo.org/secsoc/information-resources/publications-and-tools/books-and-reports/WCMS_SECSOC_20840/lang%2D%2Den/index.htm. Accessed 27 Mar 2019
United Nations International Children's Emergency Fund (UNICEF) (2012) Integrated social protection systems: enhancing equity for children. UNICEF, New York. https://www.unicef.org/socialpolicy/index_62100.html. Accessed 27 Mar 2019
Vandenhole W (2016) Is the right to social protection qualified by a duty to work? In: Parmentier S, Werdmölder H, Merrigan M (eds) Between rights and responsibilities. A fundamental debate. Intersentia, Antwerp, pp 85–106
Vandenhole W, Benedek W (2013) Extraterritorial obligations and the north-south divide. In: Langford M, Vandenhole W, Scheinin M, van Genugten W (eds) Global justice, state duties. The extraterritorial scope of economic, social, and cultural rights in international law. Cambridge University Press, Cambridge, pp 332–363
World Bank (2012) Resilience, equity, and opportunity. The World Bank Social Protection and Labor Strategy 2012–22. The World Bank, Washington DC. http://documents.worldbank.org/curated/en/443791468157506768/pdf/732350BR0CODE200doc0version0REVISED.pdf. Accessed 27 Mar 2019
World Bank (2015) Going universal: how 24 countries are implementing universal health coverage reforms from the bottom up. The World Bank, Washington DC. http://documents.worldbank.org/curated/en/936881467992465464/pdf/99455-PUB-Box393200B-OUO-9-PUBDATE-9-28-15-DOI-10-1596-978-1-4648-0610-0-EPI-210610.pdf. Accessed 27 Mar 2019
World Bank (2019) World Development Report 2019: the changing nature of work. The World Bank, Washington, DC. http://www.worldbank.org/en/publication/wdr2019. Accessed 27 Mar 2019
World Bank/International Labour Office (ILO) (2016) A shared mission for Universal Social Protection. Concept Note. www.ilo.org/wcmsp5/groups/public/@dgreports/@dcomm/documents/genericdocument/wcms_378996.pdf. Accessed 27 Mar 2019

World Bank/International Labour Office (ILO) (2018) Universal social protection. Country cases. World Bank/International Labour Office, Washington, DC/Geneva. https://www.social-protection.org/gimi/RessourcePDF.do;jsessionid=HNfANtot49wF-o5lBQb8uYIZgAiWmoF1zTLceMH1fjeHXAgIfxw1!-1894110946?id=55072. Accessed 27 Mar 2019
World Health Organization (WHO) (2010) The World Health Report. Health systems financing – the path to universal coverage. WHO, Geneva. http://apps.who.int/iris/bitstream/10665/44371/1/9789241564021_eng.pdf. Accessed 27 Mar 2019
World Health Organization (WHO) (2015) WHO Health Report 2015: from MDGs to SDGs. WHO, Geneva
World Health Organization (WHO) (2018) World Health Statistics 2018. WHO, Geneva
World Health Organization, World Bank (2017) Tracking universal health coverage: 2017 Global Monitoring Report. WHO, Geneva
Yamin EA (2008) Beyond compassion: the central role of accountability in applying a human rights framework to health. Health Hum Rights J 10(2):1–20
Yamin EA, Gloppen S (2011) Litigating health rights: can courts bring more justice to health? Harvard Law School, Cambridge
Young KG (2012) Constituting economic and social rights. Oxford University Press, Oxford

Markus Kaltenborn is Professor of Public Law at the Ruhr-University Bochum, Faculty of Law. He is Director of the Institute of Development Research and Development Policy (IEE) at Ruhr University Bochum, member of the German Institute for Human Rights (DIMR e.V.) and of the supervisory board of Oxfam Germany e.V; in recent years he has also been Visiting Lecturer at the Institute for Social Development of the University of the Western Cape (Cape Town). His main areas of research are Health Law, Social Protection Law, the Law of Development Cooperation and Human Rights Law. He is (together with Katja Bender and Christian Pfleiderer) editor of the book Social Protection in Developing Countries. Reforming Systems (Routledge 2013) as well as (together with Philipp Dann und Stefan Kadelbach) of the edited volume Entwicklung und Recht (Development and Law, Nomos 2014). Furthermore he is the author of a study on Social Rights and International Development (Springer 2015).

People and Their Health Systems: The Right to Universal Health Coverage and the SDGs in Africa

Delanyo Dovlo

Contents

Abstract The right to health is recognized as a basic human right in various United Nations official documents and in the founding principles of the World Health Organization whose constitution envisaged a right to the highest attainable standard of health for everyone. The health implications of the SDGs is linked to fundamental Human rights that the 2030 Agenda is anticipated to contribute extensively to (see footnotes 9, 10, 11).

We discuss the ability of Sub-Saharan African countries to protect the health rights of its populations given the challenges of poor economic development and significant poverty levels though some countries (Rwanda, Ethiopia, Ghana, etc.) have improved health services coverage by removing financial barriers. The right to health can be expensive and African countries did increase their health budgets, as decided in the "Abuja Declaration" target of allocating 15% of overall government budgets to health. Between 1990 and 2013 this allocation did increase from an average of 3.7% to 11.4%.

Attaining health rights in Africa requires certain policy emphases including protections from catastrophic expenditures for health, ensuring access to quality health services, and building effective "voice" for populations to exercise their rights. Enablers of health rights should include good policy and governance, with expanded social movements; and SSA countries should seize upon crises such as the

D. Dovlo (✉)
Independent Consultant, Accra, Ghana

WHO Regional Office for Africa, Brazzaville, Congo

M. Kaltenborn et al. (eds.), *Sustainable Development Goals and Human Rights*, Interdisciplinary Studies in Human Rights 5,
https://doi.org/10.1007/978-3-030-30469-0_4

Ebola outbreak to expand health rights. Scarce resources may mean rationing of health services, and it will be important to identify and utilize innovations and ICT technology that can help to make access to health care rights a reality for all.

1 Introduction

The right to health has been recognized as a basic human right in various United Nations official documents and is part of the founding principles of the World Health Organization whose constitution envisaged a right to "a state of complete physical, mental and social well-being, not merely the absence of disease" and recognizes a "right to the highest attainable standard of health" as a fundamental human right for all persons. This right to health entails the access to timely, adequate and reasonably priced health care which is still a problem for billions of poor and vulnerable peoples globally. The right to health is closely interrelated with other human rights such as the right to education, food, work, housing, non-discrimination and access to information.[1]

In September 2015, UN member states adopted unanimously the 2030 Agenda for Sustainable Development and its set of 17 Sustainable Development Goals (SDGs) with 169 targets aimed at achieving holistic economic and social sustainable development.[2,3] The 2030 Agenda and its Sustainable Development Goals are integrally linked to the human rights obligations of all UN Member States' as indicated in international human rights agreements and various international and regional instruments (see footnote 3).[4] The SDGs recognize and promote all human rights, along with gender equality and the empowerment of women and girls[5,6] and there is clearly progressive recognition that human rights are crucial to the achievement of sustainable development (see footnotes 7, 8).

The 2030 Agenda identifies human rights as being fundamental to the SDGs, and highlights the Agenda's basis in the Universal Declaration of Human Rights and international human rights treaties and therefore emphasizes the obligation of member states to respect, defend and support human rights and fundamental freedoms for all, without discrimination with respect to race, colour, sex, language, religion, political or other opinions, national and social origin, property, birth, disability or other status. It emphasizes the right to the various freedoms from non-consensual actions, and entitlement to a broad set of cares and services without discrimination. It

[1]WHO (2017a).

[2]Dattler et al. (2016).

[3]United Nations, Office of the High Commissioner for Human Rights (2015a, b).

[4]Plan International (2016).

[5]United Nations, Office of the High Commissioner for Human Rights (2015a, b).

[6]United Nations (2015b).

also emphasized the interdependency and inter-relatedness of all rights (see footnote 6).[7] Health rights will benefit from the institution of fundamental human rights everywhere.

The SDGs focus on availability, accessibility, affordability and quality of education, health, water and other services which gives practical reality to human rights.[8] Its goals and targets include access to safe, healthy and adequate food for all, universal health coverage, free equitable and quality primary and secondary education, access to safe and affordable water, sanitation, hygiene and housing, and access to "safe, effective, quality and affordable essential medicines and vaccines for all" (see footnote 7), a strong emphasis on health. The SDGs health implications and its links to fundamental human rights is therefore quite clearly stated and it is anticipated that the goals shall contribute extensively to the realization of human rights[9,10] as practical tools for development and not just as abstract ideologies.[11] Commitments to the SDGs are political, however prior human rights treaties are legally binding on the states that have ratified them and this should encourage effective advocacy action on realizing their human rights obligations.[12]

Good governance and stewardship at all levels is critical to attaining human rights, through the rule of law, democracy, access to justice and to information, transparency and accountability, and the peace and security that is essential for sustainable development. The UN's crucial role in facilitating global governance means an operational United Nations system would have an important role to play towards attainment of the sustainable development goals and its implications of assuring human rights.[13,14] The WHO as a UN Specialized agency has an important role to play in the realization of health rights as part of other human rights.

2 Health Rights Are a Challenge for Low Income Countries

The ability of Sub-Saharan Africa to protect all the rights of its populations is hampered by poor economic performance, significant poverty levels and relatively low literacy rates in many communities. In some settings, political systems may appear to undermine fundamental access to rights including health rights and with good governance considered a prerequisite for rights based social and economic systems, governance specific to the health sector in Africa, has been seen as a

[7]Fukuda-Parr (2001).

[8]The Danish Institute for Human Rights (n.d.-a).

[9]The Danish Institute for Human Rights (n.d.-b).

[10]Hunt (2015).

[11]Mclnerney-Lankford (2017).

[12]Tascioni (2016).

[13]The Center for Economic and Social Rights (n.d.).

[14]Evans (2012).

fundamental challenge to improving and delivering effective health services and sustaining health rights.[15] Coverage and access to health services has been variable across the Africa region especially among its lowest income countries though some countries (Rwanda, Ethiopia, Ghana, etc.) have significantly increased coverage and made progress towards removing financial barriers to health. Sustaining the right to health can therefore be an expensive affair and the countries of the WHO Africa region made a commitment in 1990 through the Abuja Declaration[16] that expected countries to increase average health spending as a proportion of government budgets to 15%. Between 1990 when the declaration was made and 2013, we have seen an increase from an average of 3.7% to 11.4% of government budgets being devoted to health which is significant if still below the declared target.[17] On average, Total Health Expenditure (THE) had also increased from an average of US$95 per capita in 1995 to $222 per capita in 2013[18] with the significant investments from disease specific global health initiative funds (GAVI, GFATM, PEPFAR, etc.). Average Governments' expenditure on health also rose from $40 to $110 per capita and these combined investments have led to improved life expectancy (50 to 58 years between 1990 and 2013).[19] However, these investments have significant external funding components and still need to reflect sustainability over time in order to sustain expanded access to improved health care and maintain the gains made so far.

The value of health as a basic human right is also important to the economic growth that is essential for LICs development and recent estimates by WHO-AFRO indicate an overall economic burden of ill health in the region at over US$2 trillion per annum with an economic gain of approximately $1 trillion per year in 2030 if certain SDG 3 targets are met.[20] The SDG era therefore provides an additional impetus to expand the right to health as a crucial development channel, founded on rights based principles and an overarching goal of poverty eradication, with "*no one left behind*". This approach requires cross-sectoral, interdependent and inter-disciplinary actions in order to achieve important equity targets in general and for health in particular, with "*each country having primary responsibility for its own economic and social development*".[21]

Exercising the right to health in the low income and fragile economies in Africa faces a number of challenges. Health encompasses a number of rights and entitlements that shall need to be well defined and prioritized in order to be properly assessed and monitored in terms of where countries are in assuring these rights.[22] Due to limited financial resources and other knowledge management and

[15]Yates et al. (2017).

[16]OAU (2011).

[17]WHO Regional Office for Africa (2011).

[18]WHO Regional Office for Africa (2011).

[19]WHO Regional Office for Africa (2011).

[20]WHO (2018).

[21]United Nations (2015a).

[22]Mann (2006).

institutional capacities needed to establish a sustained rights based approach to health, decision makers should first acknowledge and establish what rights are primarily at risk and what delivering these rights to populations lacking them would mean in practical terms, and then determine a sequence of steps to be taken to turn these rights into concrete services. Prioritizing what rights take precedence in a situation of limited resources may in itself become discriminatory if this leads to providing services to a majority rather than the relative few who may be more vulnerable or are part of marginalized sections of the population.

Universal Health Coverage, the critical target 8 of SDG3 and an essential measure of the right to health, requires 3 core components—namely, (i) protection from financial hazards when health care is accessed, (ii) availability and access to services of adequate quality, and (iii) a scope of services covered that deal with the main causes of illness and obstacles to wellbeing in a comprehensive way through the life-course. We discuss below how attaining universal health coverage provides a path to the realization of health rights as part of the SDGs.

- *Right to financial protection and avoiding catastrophic expenditure for health needs*

As indicated earlier, providing access to health care services in ways that enables health rights to be exercised fully is expensive. A major obstacle to the right to health for many citizens is the ability to pay to access and use services and that even when able to pay, it does not result in impoverishment that pushes families deeper into poverty. Financial barriers to receiving health care remain some of the most persistent obstacles to realizing the right to health in low income countries. Overcoming this obstacle is a major underpinning of attaining SDG 3 and its target 8 (Universal Health Coverage), in terms of providing access to services without individuals and families incurring catastrophic expenditures that push them further into poverty. Health services cost money and do need to be paid for, however, one may argue that it is incumbent on governments and society to find ways to finance health services without requiring citizens to bear full and direct costs, especially for life-saving care and care that allows a client to pursue his or her economic activities fully and contribute to society.

- *Right of access to life saving services and to responsive health services of good quality*

The WHO, through its constitutional mandate[23] assists countries to operationalize the right to health by providing technical assistance and support as needed to design and deliver an essential package of health services, and to make these services accessible to all of its population. The size, scope and distribution of these services and the ability of even the most vulnerable or marginalized populations to access services should be a measure of a country's ability to allow the effective exercise of health rights. As indicated earlier, the ability of African and other low-income

[23]WHO (2016).

countries to mobilize resources for defined essential services is dependent on how broad a spectrum of services can be made available to the entire population. Countries therefore need to define what goes into the agreed package of essential health services that each citizen should have a right to. In some cases it may mean defining which citizens (often using income levels) is entitled to receive services that are cost free or subsidized. This may translate into a "rationing of rights" that may be counterfactual to the need to provide access for all citizens. However, the practical reality in many resources constrained environments is that exercising rights gets limited by the rationing of health care services.

It is important to note however, that even when an essential package is designated, available and accessible, it may not always result in improved health and health rights if the quality of services rendered is poor and does not result in the right outcomes to the patients or populations served. Health services should also be designed and the content of care packages inclusive of interventions that can be responsive to different contexts and population needs that is usually varied even within each country's borders. Context and geographical approaches that recognize and respond to cultural and social differences are essential to making the right to health a reality across various population and ethnic groups within each country. Governments and health service providers should therefore be responsible for designing and providing services that are culturally appropriate, and can respond positively to issues of gender, cultures and stigma that create or exacerbate vulnerabilities.

- *Individuals and communities' voice and expectations of health rights and capacity to exercise rights*

Outside of the strict boundaries of the technical constituents of Universal Health Coverage are a number of other elements that facilitate the correct demand and utilization of UHC. Expressing health rights in low income and traditional indigene settings shall need understand not only how individuals' rights are exercised but also on how expression of those rights impact on other individuals and society at large. For example individuals' right to confidentiality and non-disclosure of disease conditions are a well-recognized right. However these personal rights may translate into risk of deadly communicable disease outbreak or other major risks of harm to other individuals and communities. Under these circumstances, an individual's rights may come under pressure from its competition with broader community needs. This is especially important when resources are inadequate to police and assure protection to both the individual and his/her community. In these constraining situations and during certain crises, some health rights may have to be suspended, if it can be established as this is in the public and community's interest.

Health sector decision makers need to recognize in more depth that communities are like complex organisms whereas health systems and structures tended to be inflexible and ordered in ways that can be non-responsive to the nuances of complicated societal relationships with officialdom. This in the African context implies recognizing and dealing with the ethnically diverse urban and rural groups with complicated religious and cultural belief systems that can defy modern approaches to exercising health rights. In the Sahelian countries of sub-Saharan Africa for example, enabling health rights for nomadic pastoralist groups that do not always respect borders and the usual norms of land use may generate non-compliance with health regulations, while impinging on the health rights of other ethnic groups.

In countries and communities without adequate democratization of power and influence and the capacity and space to exercise health rights, an ill-informed populace is not empowered and often lacks the civil society organizations that can represent and articulate their interests and needs effectively and generate a response. Civil society engagement experiences elsewhere have had enormous impact on expanding health rights in Brazil and other parts of South America with the emergence of social movements that allowed for a positive interaction with government to engender influence for neglected and vulnerable populations and to organize political power for the realization of rights.[24]

There is of course a reality that CSOs and other "spokespersons" for populations (e.g. parliamentarians, unionists) may not be true representatives of communities or be suborned into becoming a part of the government and other powerful influencers.

Health professionals, themselves often a backbone of the upper middle class, may well be a source of power and influence that may knowingly or unknowingly act against the best interests of the majority and the vulnerable sections of society. Where resources are scarce, the less vulnerable may tend to be left out—and what may well be a well-meaning issue of making efficient use for resources, turn otherwise into simply a giving satisfaction of the majority and the influential components of society.

It is therefore critical that the right to health is also articulated in terms of the "*public good*" it offers to all classes and sections of society as well as the entire population. Communicable diseases that may have been initially incubated in a marginalized sub-group can quickly affect the whole country through outbreaks that put people and economies at risk (e.g. the Ebola crisis in West Africa). Other diseases of poverty (both communicable and non-communicable) can reduce productivity and place unacceptable costs on the health expenditures and budgets.

Analyzing and understanding the positive influences that the right to health brings to individuals, communities and countries as a whole is essential to driving policy that results in turning these rights into practical reality and into clear health results.

3 Facilitators and Enablers for Mainstreaming Health Rights in Africa

Our observation in sub-Saharan Africa is that for health rights to become embedded in governance and policy, decision makers must identify and enhance the effect of certain critical enablers and facilitators of health and human rights. This involves

[24]Fleury (2011).

engaging with critical stakeholders to understand and acknowledge health human rights and its public good benefits and provide the leverage to apply this awareness and knowledge to policy making.

Facilitating rights is dependent on a number of factors experienced in the Africa Region of WHO. Managing rights requires engagement with complex and interrelated factors in order to build a sustained process towards applying human rights.

- *Public policy and governance capacity and environment as an enabler*

LICs, especially the most fragile, often lack the research, analytical and institutional capacity to inform and influence policy development and decision making in ways that lead to the best possible rights results. The presence of significant donor resources in health in low income country settings means that analysis that is available often comes from donor and external sources which may at times lack the insights into the influences that affect governance effectiveness—(e.g. ethnicity considerations, and corrupt practices) that may hide real sources of power and influence that undermines health rights. In many of the region's countries, the needs and actions to address rights may vary in different parts of the same country and between countries, with varying social or religious norms that drive how these different contexts accept or adopt health policies and rights. The clarity of a right as a positive in one community may not be as lucid as an advantage in others and in situations of centralized governance, this may result in non-compliance with public health laws and other actions that are aimed at facilitating the rights to health.

- *Building Population and communities awareness and influence as an enabler of rights*

Linked in complex ways to the governance issue is the influence that traditions and culture can have on delivery of public goods. In simple terms, communities and population groups may perceive attempts to expand rights to health as impositions that may be detrimental to a way of life. In some parts of Africa, this may occur with groups that have practices like female genital cutting, various traditional religious societies, and in male dominant cultures that may undervalue girl children and their education, as well as various religious and cultural norms. The practices by certain cults in Sierra Leone significantly influenced and resisted Ebola prevention interventions during the West Africa outbreak and may have facilitated aspects of the worst EVD outbreak in history.

Therefore a question that can be posed to all health sectors is, "how should health public goods and rights be respected while also respecting cultures and traditions of individuals & communities?" secondly, "How can we avoid engendering gender discrimination, ethnic marginalization and stigma within vulnerable groups by being responsive to local cultural factors?"

- *Encouraging social movements and leadership*

Exercising health and other social rights is usually achieved through major social action against various entrenched interests and influences. Most indigenous and poor communities are unable to generate the advocacy, power and partnerships needed to achieve these aims. Communities are often made to see rights as a privilege to be granted by governments and their leadership rather than as entitlements of citizenship. Establishing the right to health can only be sustained if linked to the other social rights of the most marginalized communities. It must be part of building political will and be essential for gaining political power (in democratic contexts) and for influencing other government outcomes (e.g. mining or land use rights) in order to achieve health rights.

The community leadership capacity required in achieving these rights and objectives are often missing or lack appropriate authority to exercise influence in positive directions in modern world systems. The interest of leadership in many low resources settings may also be geared towards protecting ethnic interests and political reality rather than through alliances built around class and economic interests. These leadership capacities may need to be carefully nurtured and matured as necessary facilitators of health and other human rights.

- *Use critical events and crises to foster health rights*

A crisis, especially a health one, can provide an opportunity for change and can be used to influence progress on the right to access health services. It is said that the 2nd world war created the opportunity for the UK to establish the NHS to provide health rights to the returning soldiers and their families and as part of a prize for the common sacrifice made by all classes of society for victory in that war.[25]

The existential challenge that Ebola posed to the three West African countries that were most severely affected may have concentrated minds of politicians and decision makers on the importance of ensuring that citizens had access to robust health services that provide protection not only to local communities, but is also an important hedge for continued to economic development and societal gain. In the case of Ebola, the international community also clearly found a need to enable health (and other) rights as a mutually beneficial public good for even in the poorest countries.

The HIV/AIDS pandemic was another example of how a health and social crisis had significantly transformed health care and rights, for example in certain situations, removing what seems to be a stranglehold by health professions over certain treatments and care roles and which allowed for a delegation of tasks to less expensive cadres of health workers in order to meet the health needs of the populations at risk. The HIV crisis also provided a platform to debate and arrives at the understanding that preventive action was not enough to mitigate the epidemic but the need to provide treatment was also as important to fighting the scourge and assuring the extension of health rights to all kinds of population groups that were previously unrecognized or marginalized through stigma.

Similar lessons came up again during the Ebola crises, ensuring that establishing treatment centres and offering relief to affected individuals and communities

---

[25]Gorsky (2008).

including those in quarantines was as important as all other public health control measures. Public Health and social actions on their own would not have led to the end of the outbreak without innovations in engaging with communities and designing the outbreak response in ways that respected local culture and gained confidence of various population groups with the efforts to stop the outbreak. It is important that the international community as well as governments and CSOs be prepared to confront health crises and utilize the advocacy and innovation that such incidents bring to create better services for their populations' needs and to provide access to the expected health rights.

- *Education and fostering gender parity as a facilitator health rights*

Certain Sub-Saharan Africa cultural contexts did not give women full status to their health rights and this often created detrimental social, cultural and sexual right effects. These effects can clearly be moderated by education and economic empowerment of women which can also enhance women's connection to the health rights of their children and families. Given the right conditions, women in most communities are well placed to exercise and utilize health services effectively and engaging women as a centerpiece in exercising health rights and utilization of health services can improve how communities demand and express the right to health. The health of women and families, perhaps best measured through maternal mortality ratios, is one of the major indicators of how the right to life and to health is manifested in a country. The disparity in maternal mortality ratios between countries rich and poor, demonstrates an unacceptable level of avoidable death that clearly indicates the multiple obstacles and challenges undermining access to services that allows for a full expression of the right to safe child birth and care. The WHO Global Observatory[26] indicates for example maternal mortality ratio estimates for Equatorial Guinea at 342 deaths per 100,000 births, and for Botswana 129/100,000 even though Gross National Income (GNI) per capita is relatively close at PPP$7180 and $6750 respectively. In Sierra Leone, WHO maternal mortality estimates indicate that annual death rates may be even higher than the total deaths that occurred during the Ebola outbreak in that country comparing to figures seen in the year 2000.[27] Moreover, the outpouring of support and resources observed in the response to Ebola, has not been translated at anywhere near the same level, with the country's maternal mortality problem.

Operationalizing the right to health in fragile and low income countries will require effective tackling of several challenges, some of which are discussed in the previous sections. The economic and social situation in the WHO's Africa region present obstacles to the full exercising of health rights by individuals and societies and even governments willing to provide access to these rights can be stymied by these factors. The right to health in low income countries is linked to the right to universal health coverage as set of defined services and interventions that are within

[26]WHO, Global Observatory Database (n.d.-b).

[27]Figueroa et al. (2017).

the resources and need of a country to provide. This right is linked to a number of other important rights (education, water and sanitation, etc.) that are important for human dignity.

- *Should one right be prioritized over another?*

This brings up the issue often confronted in health and raised by the HIV pandemic for example—can one health right be prioritized over another? How can this prioritization be made in a rational way? Are there priority rights exercised within the health sector and how do these match with other social rights that are also priorities in resources constrained contexts?

The challenge governments may face is having a hierarchy if rights which may mean suppression of other rights and elevation of others which may mean different benefits for different populations. For example, a government may prioritize Malaria services affecting 50% of the population over lymphatic filariasis affecting less than 1% but scattered and inaccessible. Each has debilitating social and economic effects on the individuals and communities affected but which should be tackled first? It is further exacerbated as a rights challenge when the prevalence of one condition or the other affects marginalized or vulnerable groups and populations who may also be denied other rights but lack the political influence to get remedies.

The idea of prioritizing rights in health requires agreeing on a hierarchy of needs and determining as part of a country's social contract, what rights can be demanded? What will be the criteria for deciding what is fundamental to each population group in a country and how can resources be designated to meet these needs in ways that achieve practical health outcomes? What should be available, even if not demanded as a right by the population?

In our work in Africa, a critical need is simply the protection from impoverishment or catastrophic expenditure when accessing health care. The data from the WHO-AFRO health financing shows that quite a significant number of the 47 countries in the Africa WHO region have out-of-pocket expenditures on health constituting above 40% of total health expenditure[28] a limit consider likely to spur catastrophic expenditure for poor families or likely to result in impoverishment. One may argue that it is a primary right of citizens not to have to choose between impoverishment and health care, and it is incumbent on governments and communities to find ways to ensure that money does not obstruct the right to health.

- *Right of access to responsive services of good quality*

Countries that offer access to health services should also guarantee that these services are the right services that will provide the right results, and that these are delivered in ways that are convenient and impactful for individuals and communities. This is again a factor in the ability to fully exercise and benefit from the right to good health. A number of factors undermine this aspect of facilitating rights. In most LICs, trained and qualified health workers are few in number and unevenly

[28]WHO Regional Office for Africa (2011).

distributed across a country.[29] The internationally mobile cadres are often the subject of attraction and migration to richer countries, despite the investments made by their own countries to train them. Receiving countries consider it the right of the well-trained individual (but not the unskilled) to be able to migrate freely and work where they chose to. International unregulated migration of health professionals from poor to rich countries therefore undermines the right to health for populations in the source countries. Countries may have to make the choice between the individual rights of health professionals they've heavily invested in, and their ability to enable rights to health of the wider population.

4 How Can Africa Realize the Right to Health as Part of the SDGs?

SDG 3 is the health SDG that expresses a "healthy lives and wellbeing" goal as part of a sustainable development ideal, contributed to, but also contributing to many other SDGs. To much of the health sectors global operators, target 8 in Goal 3 (achieving Universal Health Coverage) best expresses an effort to realize the right to health for all populations. Practical steps are needed as many of the issues identified above as affecting or influencing health rights outcomes, require significant resources and capacity to fulfill needs. The WHO's and other health stakeholders efforts to achieve the SDGs are ways of actualizing the right to health and gradually expending its coverage and effectiveness to entire populations often under circumstances of scarce or misapplied resources and ineffective health strategies.

However, Africa has been a major beneficiary of the Global Health Initiatives such as GAVI, GFATM, PEPFAR, etc. that have mobilized significant resources to fight certain major diseases often in ways that do not build sustained health systems and informed decision making. In the absence of good drinking water and sanitation, and the absence of food, are the beneficiaries of excellent vaccination programs and treatment schedules likely to die from conditions that are even less expensive to resolve? The availability of funding for global priority programs generated some very good results but also some health systems challenges especially where absorptive capacity for resources was low and not strengthened prior to receiving these sizeable resources.

- *Governance, leadership and accountability weaknesses* at national and local levels

Many LICs and indeed MICs in the Africa region are constrained by overall leadership and governance weaknesses that undermine achievement of results and

[29] Buchan et al. (2013).

being held accountable to the population for resources and availability and effectiveness of health services. There is an often high component of external donor resources in the Total Health Expenditures (THE) of most countries in Africa, linked to various "conditionalities" aimed at ensuring certain governmental actions. Sometimes, these conditionalities mean that government officials tended to owe more accountability to the sources of external funding than to their tax paying populations. Conditionalities may not always be a bad thing as they may include requirements for governments to for example provide services for marginalized groups or to implement gender sensitive actions.

However domestic resources, even in many low income countries, do constitute the higher proportion of the THE.[30] Reviews of total health expenditure by sub-Saharan African countries from WHO's database from National Health Accounts, indicates that some 60–66% of THE is from internal resources with about 30% from external resources and donors.[31] The lack of accountability for mobilizing and effectively utilizing these resources may undermine Universal Health Coverage (UHC) and the right to health. Improving governance and accountability in its broadest sense at all levels, including in districts and at community levels should be essential ingredients to the exercise of health rights and to manage the gradual fulfillment of UHC, the right to health and other human rights.

It is quite clear that healthy lives and wellbeing, as ordained in the SDGs, can only be achieved with extensive action that is outside of the health sectors. Which is why a broad multi-sectoral engagement on human rights will be critical to broadening and deepening health rights and wellbeing, which is defined by WHO as not only the absence of disease, but presence of physical, social and mental wellbeing.

The effort to bring about rights, especially to marginalized populations can only be effective if it involves a holistic approach that encourages access to all human rights and not only to health rights. The fragmentation in approach between various sectors that sometimes affects government agencies effectiveness can only undermine the ability of individuals and communities to exercise their rights.

An EU WHO Program on UHC that currently also involves bilaterally Luxembourg and Ireland, targets a number of low income countries in the region and elsewhere with a focus on facilitating policy dialogue and stakeholder consultations that encourage policy decisions and planning for universal health coverage.[32] Health systems experts are placed in the target countries as catalysts for discussion and debates on how policy can drive UHC and facilitates a rights-based approach to UHC. The results so far have been positive, moving countries on a sustained path towards effective policies that work towards attaining UHC and reaching the most vulnerable in society.

---

[30]Elovainio and Evans (2013).

[31]WHO (n.d.-a).

[32]Reinicke (2016).

Moreover, recent estimates of the economic impact of ill-health in the WHO Africa region indicate that non-communicable diseases (NCDs), injuries and accidents constituted about 50% of the costs to countries but these receive only minimal donor support.[33] The package of essential services that countries develop as the basic access to health rights often has limited content for NCDs, and unlike communicable diseases, there are no agreements to moderate the costs of medicines and laboratory testing, etc. in order to effectively manage these diseases that have such a high impact on economic development.

This is why mobilizing and utilizing domestic resources effectively should become an important part of ensuring that priorities reflecting the actual health needs and rights of local populations are tackled and not suppressed and replaced completely by the global push to prevent spread of communicable and outbreak prone diseases.

An important aspect of WHO's work on improving health in the region is the Africa Health Observatory, established to coordinate a network of national health observatories that seek and publish improved data and information, and expands analysis and evidence capacity to encourage its use in policy dialogue and policy making. Information and evidence is essential to guide countries towards utilizing effective interventions and basing their actions on proven methods.

Data and evidence should play a major role in mobilizing communities and decision makers for health rights. An important aspect is to scale up new electronic and mobile communication technologies with quite high uptake rates for mobile phone technology in the region. The WHO entered into an agreement with the International Telecommunication Union in October 2017, to collaborate on expanding access to these technologies in the health sector.[34] The use of these technologies for enhancing data and evidence availability is important for understanding trends and impact but they also offer possibilities for expanding access to services in remote and hard to reach areas through telemedicine and other forms of virtual consultations and treatment.

- "*Thinking inside the box*" and sustained implementation of initiatives and ideas

The Africa region over the past 2–3 decades have been flooded with numerous initiatives and innovations, often originating from donor or partner countries and backed by funding, which make it very difficult to resist. These initiatives are often not informed by in-country expertise and tend to be experimental and short term. This does not allow for a sufficiently graduated learning curve for countries to internalize the approaches used and thus build sustainability.

The terms "low hanging fruit", "innovation" and "think outside the box" are often applied to a plethora of health innovations and initiatives though they are often

[33] Nugent and Feigl (2010).

[34] WHO (2017b).

difficult to implement fully and ascertain benefits. . There is constant pressure to try something new, especially when backed with resources, but without enough efforts to build on prior experiences and existing interventions and to gradually scale these up. Universal access to health rights will require sustained effort to solve problems and learn lessons from implementation processes.

Investments from the global funds and other donor arrangements should also focus on building strong institutional research and analytical capacity, that is stable and not fleeting and allow for serious efforts to exhaust "inside the box" opportunities, before jumping to the next "innovative" idea. Incremental changes should be the required model with due regard to country contexts with gradual building of capacity and knowledge for sustained effect.

- Local and context specific responses and ownership of solutions

There is a need for health rights implementation solutions to be locally based. Globally determined strategies and approaches need to be significantly contextualized to reflect how local cultures and norms absorb and utilize health rights. A plethora of initiatives and one-off research and pilots don't get grounded into full reality and while there may be useful contributions to international academic knowledge and research, they often do not meet the sustainability test of the SDGs in establishing and sustaining health rights. This can be a difficult undertaking as communities in each country can be complex and multifaceted organisms, often with significant shifts and changes over time that require an evolving response to needs and demands. Therefore rights development efforts in health need to be sustained over time to build the capacity and confidence of communities' stakeholders and individuals and in order to achieve a well-grounded understanding of health and the rights to it.

- Closing "*knowledge & awareness gap*" between services and their communities

Accessing the right to health requires more effort at narrowing the gap in knowledge between communities and their governments and providers; both on how services should be organized and delivered but also on what their rights are and what is legally required of elected officials. It is essential to mobilize capacity for rights empowerment by creating "facilitators" for operationalizing health rights and moderating the interactions between communities and health systems managers as a way of building communities and individuals' capacities to exercise their human rights in health. Without populations and communities' internalization and ownership of their rights to health and their investment of effort to realize rights, universal health coverage remains a mirage.

5 Conclusion

The strong focus of the SDGs on equity with the slogan of "*no one left behind*" provides a good platform for LICs to give practical reality to the right to health and to improving the lives of most disadvantaged in countries and communities. In order to achieve these, each country must be held accountable on their primary responsibility for the population's own economic and social development and the overarching goal of poverty eradication.

The effort towards universal health rights will require significant cross-sectoral action and partnerships to achieve measurable results.

In order for everyone including the most vulnerable to exercise their right to health, services must be designed based on the resources available, to be culturally appropriate and to mitigate any negative impact due to gender, stigma and other vulnerabilities.

Improved governance and accountability with strong institutions that function well will be crucial to sustaining all human rights including the right to health.

As countries build effort towards realizing the SDGs and Goal 3, a clear vision needs to be articulated on how these efforts can lead to realization of the right to health. This, it is proposed will require the following strengths to be boosted.

1. Countries should build strong sub-national and local governance systems with efficient primary health care systems that bring accountability, resilience and security closer to each population group.
2. Improved governance processes including evidence based policy dialogue, that should lead to strategic prioritization of critical services that expand rights to health and remove financial, geographical & social barriers.
3. Build and/or strengthen institutions and processes that oversee and monitor the obligation of countries under the Universal Declaration of Human Rights and various related rights treaties and covenants of the UN, to respect, defend and support human rights and fundamental freedoms without discrimination and also the entitlement to certain sets of cares and services including for health.
4. The increasing role of ICT in democratization and communication of basic and other rights provides an opportunity that should be expanded in Africa to improve access to health care and facilitate the monitoring of health care trends and coverage.

The above, along with other factors should constitute the facilitators and enablers of health rights that allow for the demand for rights to develop as well as the accountability for providing access to health care and rights.

Analyzing and understanding the positive influences that the right to health brings to individuals, communities and countries as a whole is essential to driving policy that results in turning these rights into practical reality and into clear health results.

References

Buchan J, Couper I, Tangcharoensathien V, Thepannya K et al (2013) Bull World Health Organ 91:834–840. http://www.who.int/bulletin/volumes/91/11/13-119008/en/. Accessed 02 Jan 2018

Dattler R, Barclay D, Marshall A (2016) Sustainable development goals and human rights. The International Planned Parenthood Federation, June 2016. https://www.ippf.org/sites/default/files/2016-11/SDG%2BHR_facts.pdf. Accessed 02 Jan 2018

Elovainio R, Evans D, Raising and Spending Domestic Money (2013) Chatham House, Working Group on Financing Paper 2, May 2013. https://www.chathamhouse.org/sites/files/chathamhouse/public/Research/Global%20Health/0513_healthfinance.pdf. Accessed 03 Jan 2018

Evans A (2012) The UN's role on sustainable development. New York University: Center on International Cooperation. http://cic.nyu.edu/sites/default/files/evans_sustainable_development.pdf. Accessed 28 Dec 2017

Figueroa C, Linhart C, Beckley W, Pardosi J (2017) Maternal mortality in Sierra Leone: from civil war to Ebola and the sustainable development goals. Int J Public Health. https://link.springer.com/content/pdf/10.1007%2Fs00038-017-1061-7.pdf. Accessed 02 Jan 2018

Fleury S (2011) Brazil's health-care reform: social movements and civil society. Lancet 377 (9779):1724–1725. https://www.ncbi.nlm.nih.gov/pubmed/21561650. Accessed 31 Dec 2017

Fukuda-Parr S (2001) Indicators on human development and human rights-overlaps, differences ... and what about the human development index? Stat J United Nations ECE 18:239–248. http://sakikofukudaparr.net/wp-content/uploads/2016/06/Indicators-of-human-development-and-human-rights.pdf. Accessed 02 Jan 2018

Gorsky M (2008) The British National Health Service 1948–2008: a review of the historiography. Soc Hist Med 21(3):437–460. https://doi.org/10.1093/shm/hkn064. Accessed 02 Jan 2018

Hunt P (2015) SDG series: SDGs and the importance of formal independent review: an opportunity for health to lead the way. Health Hum Rights J, 2 September 2015. https://www.hhrjournal.org/2015/09/sdg-series-sdgs-and-the-importance-of-formal-independent-review-an-opportunity-for-health-to-lead-the-way/. Accessed 02 Jan 2018

Mann J (2006) Health and human rights: if not now, when? Am J Public Health 96(11):1940–1943. https://ajph.aphapublications.org/doi/pdf/10.2105/AJPH.96.11.1940. Accessed 31 Dec 2017

Mclnerney-Lankford S (2017) Human rights and the SDGs: progress or a missed opportunity?. Oxford Human Rights Hub. http://ohrh.law.ox.ac.uk/human-rights-and-the-sdgs-progress-or-a-missed-opportunity/. Accessed 02 Jan 2018

Nugent RA, Feigl AB (2010) Where have all the donors gone? Scarce donor funding for Non-Communicable Diseases. Center for Global Health Development, Working Paper 228, November 2010. http://www.cgdev.org/content/publications/detail/1424546. Accessed 02 Jan 2018

OAU (2011) Abuja declaration of HIV/AIDS, tuberculosis and other related infectious diseases. Organisation of African Unity, OAU/SPS/ABUJA/3. http://www.un.org/ga/aids/pdf/abuja_declaration.pdf. Accessed 30 May 2018

Plan International (2016) Making the link: SDGs and human rights obligation. https://plan-international.org/publications/making-link-sdgs-and-human-rights-obligations. Accessed 02 Jan 2018

Reinicke M (2016) The EU-Lux-WHO-UHC partnership: a 'cutting-edge' programme - a view from the European Union. Universal Health Coverage Partnership, 23 May 2016. http://uhcpartnership.net/the-eu-lux-who-uhc-partnership-a-cutting-edge-programme-a-view-from-the-european-union/. Accessed 03 Jan 2018

Tascioni G (2016) SDGs and human rights: how to measure States' compliance? Eur J Sustain Dev 5(3):140–150. https://ecsdev.org/ojs/index.php/ejsd/article/view/340/337. Accessed 29 Dec 2017

The Center for Economic and Social Rights (n.d.) Human rights in sustainable development. http://www.cesr.org/human-rights-sustainable-development. Accessed 02 Jan 2018

The Danish Institute for Human Rights (n.d.-a) The human rights guide to the sustainable development goals. The Danish Institute for Human Rights. http://sdg.humanrights.dk/. Accessed 02 Jan 2018

The Danish Institute for Human Rights (n.d.-b) Tying human rights and sustainable development goals together in a mutually reinforcing way. https://www.unece.org/fileadmin/DAM/env/pp/wgp/WGP-20/Statements_and_Presentations/Danish_Institute_for_Human_Rights_Birgitte_Feining.pdf. Accessed 03 Jan 2018

United Nations (2015a) Addis Ababa action agenda of the third International conference on financing for development: Addis Ababa action agenda. United Nations, 27 July 2015. http://www.un.org/esa/ffd/wp-content/uploads/2015/08/AAAA_Outcome.pdf. Accessed 02 Jan 2018, p 5

United Nations (2015b) Transforming our world: the 2030 agenda for Sustainable Development. http://www.un.org/ga/search/view_doc.asp?symbol=A/RES/70/1&Lang=E. Accessed 04 Jan 2018

United Nations, Office of the High Commissioner and Human Rights (2015a) Human Rights and the 2030 Agenda for Sustainable Development. http://www.ohchr.org/EN/Issues/MDG/pages/The2030Agenda.aspx. Accessed 03 Jan 2018

United Nations, Office of the High Commissioner and Human Rights (2015b) Transforming our world: human rights in the 2030 agenda on the Sustainable Development. http://www.ohchr.org/Documents/Issues/MDGs/Post2015/HRAndPost2015.pdf. Accessed 03 Jan 2018

WHO (2016) The Global Guardian of Public Health, May 2016. http://www.un.org/esa/ffd/wpncontent/uploads/2015/08/AAAA_Outcome.pdf. Accessed 31 Dec 2017

WHO (2017a) Human rights and Health, A Factsheet, December 2017. http://www.who.int/mediacentre/factsheets/fs323/en/. Accessed 21 Dec 2017

WHO (2017b) WHO and ITU to use digital technology to strengthen public health services in Africa, 26 October 2017. http://www.afro.who.int/news/who-and-itu-use-digital-technology-strengthen-public-health-services-africa. Accessed 02 Jan 2018

WHO (2018) The productivity cost of illness in Africa. WHO Regional Office for Africa, Brazzaville. Licence: CC BY-NC-SA 3.0 IGO. https://www.afro.who.int/sites/default/files/2019-03/Productivity%20cost%20of%20illness%202019-03-21.pdf

WHO (n.d.-a) Global Health Expenditure Database. http://apps.who.int/nha/database/Select/Indicators/en. Accessed 03 Jan 2018

WHO (n.d.-b) Global Observatory Database. http://gamapserver.who.int/gho/interactive_charts/mdg5_mm/atlas.html. Accessed 30 May 2018

WHO Regional Office for Africa (2011) Atlas of African Health Statistics 2016-Health situation analysis in the African Region. https://www.aho.afro.who.int/en/publication/5266/atlas-african-health-statistics-2016-health-situation-analysis-african-region. Accessed 02 Jan 2018

Yates R, Brookes T, Whitaker E (2017) Hospital detentions for non-payment of fees: a denial of rights and dignity. Chatham House, Research paper. https://www.chathamhouse.org/sites/files/chathamhouse/publications/research/2017-12-06-hospital-detentions-non-payment-yates-brookes-whitaker.pdf. Accessed 02 Jan 2018

Delanyo Dovlo was the Director of Health Systems and Services Cluster in the WHO Regional Office for Africa (AFRO). Previously he served as WHO Country Representative in Rwanda and Health Systems Adviser at WHO headquarters. A Ghanaian public health physician with over 30 years of clinical and public health experience, he has been a consultant to various countries on Health Systems and Health Sector Reforms. He has an MB CHB from the University of Ghana, an MPH from University of Leeds, a Membership of the West Africa College of Physicians and is a Fellow of the Ghana College of Physicians & Surgeons.

Freedom from Violence, Full Access to Resources, Equal Participation, and Empowerment: The Relevance of CEDAW for the Implementation of the SDGs

Beate Rudolf

Contents

Abstract The 2030 Agenda acknowledges the key role of gender equality and the empowerment of women for the implementation of the Sustainable Development Goals (SDGs) through Goal 5 as well as through the commitment to mainstreaming gender throughout all goals and in the implementation of the Agenda. The United Nations Convention on the Elimination of All Forms of Discrimination against Women (CEDAW) is the core international human rights treaty on women's equality in all fields, and it has produced a wealth of information on causes of discrimination against women, on gaps in implementing women's human rights that prevent their full and equal participation in all areas of life as well as on successful strategies and instruments to address the structural causes of gender-based discrimination. This article examines how the monitoring processes on CEDAW implementation can be

I would like to thank Julia Kercher, former Researcher and Policy Advisor at the German Institute for Human Rights, for her excellent preparation of the oral presentation at the Conference, which formed the starting point for the present contribution.

B. Rudolf (✉)
German Institute for Human Rights, Berlin, Germany
e-mail: info@institut-fuer-menschenrechte.de

M. Kaltenborn et al. (eds.), *Sustainable Development Goals and Human Rights*, Interdisciplinary Studies in Human Rights 5,
https://doi.org/10.1007/978-3-030-30469-0_5

used for promoting gender-sensitive SDG implementation. It also analyzes the possible synergies between the SDGs and CEDAW, in particular with respect to the national, regional, and global follow-up and review processes under the SDGs. As an example, the article looks into the role of National Human Rights Institutions in this regards.

1 Introduction

Thanks to UN Women and women's organizations from around the world, women's equality, especially their freedom from violence, their full access to resources, and equal participation in decision-making, as well as women's empowerment are part and parcel of the 2030 Agenda for Sustainable Development.[1] From an early stage of the negotiations, UN Women had advocated for a stand-alone goal on women's equality.[2] The aim was to have a goal that would go beyond the Gender Equality Goal in the Millennium Development Goals[3] by addressing the structural causes of inequality. The result of the joint efforts is Goal 5 of the Sustainable Development Goals (SDGs), which reads:

> **Goal 5. Achieve Gender Equality and Empower All Women and Girls**
> 5.1 End all forms of discrimination against all women and girls everywhere
>
> 5.2 Eliminate all forms of violence against all women and girls in the public and private spheres, including trafficking and sexual and other types of exploitation
>
> 5.3 Eliminate all harmful practices, such as child, early and forced marriage and female genital mutilation
>
> 5.4 Recognize and value unpaid care and domestic work through the provision of public services, infrastructure and social protection policies and the promotion of shared responsibility within the household and the family as nationally appropriate
>
> 5.5 Ensure women's full and effective participation and equal opportunities for leadership at all levels of decision-making in political, economic and public life

[1] For the role of women's organizations, see Sen (2019).

[2] UN Women, A Transformative Stand-alone Goal on Achieving Gender Equality, Women's Rights and Women's Empowerment: Imperatives and Key Components, New York 2013, http://www.unwomen.org/-/media/headquarters/attachments/sections/library/publications/2013/10/unwomen_post2015_positionpaper_english_final_web%20pdf.pdf?la=en&vs=1454 (last accessed 15 July 2019).

[3] UN General Assembly Resolution "2015 World Summit Outcome," A/RES/60/1 (16 September 2005), paras. 58–59, later termed "Goal 3."

> 5.6 Ensure universal access to sexual and reproductive health and reproductive rights as agreed in accordance with the Programme of Action of the International Conference on Population and Development and the Beijing Platform for Action and the outcome documents of their review conferences
>
> 5.a Undertake reforms to give women equal rights to economic resources, as well as access to ownership and control over land and other forms of property, financial services, inheritance and natural resources, in accordance with national laws
>
> 5.b Enhance the use of enabling technology, in particular information and communications technology, to promote the empowerment of women
>
> 5.c Adopt and strengthen sound policies and enforceable legislation for the promotion of gender equality and the empowerment of all women and girls at all levels".[4]

Achieving women's full equality in all areas is also the object and purpose of the United Nations Convention on the Elimination of All Forms of Discrimination against Women (CEDAW), adopted in 1979. In this human rights treaty, often called the "Women's Rights Convention," the states acknowledged that discrimination against women is rooted in traditional gendered roles, and that discrimination holds women back from full participation in the political, social, economic, and cultural life of their countries and thus prevents them from bringing their potentialities to the service of their countries and humanity.[5] Given the broad ratification of CEDAW[6] and the vast experience of the CEDAW Committee, exploring the possible synergies between the SDGs and CEDAW is an idea that suggests itself so as to ensure that lessons learnt and information gathered under the Women's Rights Convention are made available for the implementation of the SDGs.

2 Women's Equality and Women's Empowerment in the 2030 Agenda

Women and their rights to equality, especially with respect to access to resources, equal public, economic, and social participation, as well as women's empowerment are not only one element of the Sustainable Development Goals (SDGs), but have an important place in the 2030 Agenda. In addition to gender equality and the empowerment of all women and girls being one of the 17 goals, many of the targets under

---

[4]UN General Assembly Resolution "Transforming our world: the 2030 Agenda for Sustainable Development", A/RES/70/1 (25 September 2015).

[5]CEDAW, Preamble paras. 14 and 7, respectively.

[6]As of 15 July 2019, 189 states had ratified CEDAW.

the other 16 goals specifically relate to the situation of women and girls.[7] Moreover, the SDGs require systematic mainstreaming of a gender perspective in the implementation of the 2030 Agenda.[8] All three elements are brought together in paragraph 20 of UN General Assembly Resolution 70/1:

> Realizing gender equality and the empowerment of women and girls will make a crucial contribution to progress across all the Goals and targets. The achievement of full human potential and of sustainable development is not possible if one half of humanity continues to be denied its full human rights and opportunities. Women and girls must enjoy equal access to quality education, economic resources and political participation as well as equal opportunities with men and boys for employment, leadership and decision-making at all levels. We will work for a significant increase in investments to close the gender gap and strengthen support for institutions in relation to gender equality and the empowerment of women at the global, regional and national levels. All forms of discrimination and violence against women and girls will be eliminated, including through the engagement of men and boys. The systematic mainstreaming of a gender perspective in the implementation of the Agenda is crucial.[9]

Consequently, the 2030 Agenda also provides that follow-up and review processes at all levels must be gender-sensitive.[10]

The reasons for the important place of women's equality in the SDGs are obvious. It is the present state of the world, pointedly described by the General Assembly: "Billions of our citizens continue to live in poverty and are denied a life of dignity. There are rising inequalities within and among countries. There are enormous disparities of opportunity, wealth and power. Gender inequality remains a key challenge."[11] Indeed, as the 2015 World Women's Report of the United Nations shows, considerable gender gaps to the disadvantage of women persist in areas such

[7]In particular: Target 2.2 (end hunger through ensuring secure and equal access to land for women), Target 4.3 (ensure inclusive quality education through equal access for all women and men to affordable and quality technical, vocational and tertiary education), Target 6.2 (ensure sanitation for all, paying special attention to the needs of women and girls), Target 8.5 (achieve full and productive employment and decent work for all women and men), Target 8.8 (protect labour rights, in particular for migrant women), Targets 11.2 and 11.7 (make cities inclusive and safe by expanding public transportation with special attention to the needs of women, and provide universal access to safe, inclusive and accessible, green and public spaces, in particular for women), and Target 13.b (combat climate change by raising capacity for effective climate change-related planning in LDCs, including focusing on women).

[8]UN General Assembly Resolution "Transforming our world: the 2030 Agenda for Sustainable Development", A/RES/70/1 (25 September 2015), para. 20.

[9]UN General Assembly Resolution "Transforming our world: the 2030 Agenda for Sustainable Development", A/RES/70/1 (25 September 2015), para. 20.

[10]UN General Assembly Resolution "Transforming our world: the 2030 Agenda for Sustainable Development", A/RES/70/1 (25 September 2015), para. 74 (e).

[11]UN General Assembly Resolution "Transforming our world: the 2030 Agenda for Sustainable Development", A/RES/70/1 (25 September 2015), para. 14.

as education, the labour market, as well as political power and decision-making.[12] The gender pay gap and gendered family responsibilities contribute largely to gender disparities in poverty, which in turn is exacerbated if women are single and with dependent children.[13] The gender gaps are all the more striking as men outnumber women worldwide by 62 million.[14] All over the world, there are considerable differences between men and women in health trajectories throughout the life cycle, due to biological factors as well as gender-based inequality and gender norms. In the Global South, particular health risks exist for women in relation to pregnancy, childbirth, and sexually transmitted diseases, caused by underdeveloped health systems as well as gender-based obstacles.[15] Violence against women is prevalent everywhere, with one third of the women worldwide having experiences physical and/or sexual violence by an intimate partner or sexual violence by a non-partner during their lives.[16]

For these reasons, the UN member states rightly call gender equality as being crucial to the realization of the SDGs.[17] Only through realizing the three-pronged approach—a separate goal, a commitment to gender mainstreaming in the implementation of the SDGs, and gender-sensitive follow-up and review—can the UN member states fulfil their pledge that "no one will be left behind"[18]—especially not women who make up half of the world's population and who are even more disadvantaged when their gender-based discrimination intersects with discrimination based on other grounds, such as belonging to an (ethnic, religious, language or sexual) minority or indigenous group, or because of their age, marital status, disability or migration background.

[12]UN Department of Economic and Social Affairs, The World's Women: Trends and Statistics, 2015, ST/ESA/STAT/SER.K/20 (2015), https://unstats.un.org/unsd/gender/downloads/WorldsWomen2015_report.pdf (last accessed 15 July 2019), pp. xi–xii.

[13]UN Department of Economic and Social Affairs, The World's Women: Trends and Statistics, 2015, ST/ESA/STAT/SER.K/20 (2015), https://unstats.un.org/unsd/gender/downloads/WorldsWomen2015_report.pdf (last accessed 15 July 2019), p. xiv.

[14]UN Department of Economic and Social Affairs, The World's Women: Trends and Statistics, 2015, ST/ESA/STAT/SER.K/20 (2015), https://unstats.un.org/unsd/gender/downloads/WorldsWomen2015_report.pdf (last accessed 15 July 2019), p. ix.

[15]UN Department of Economic and Social Affairs, The World's Women: Trends and Statistics, 2015, ST/ESA/STAT/SER.K/20 (2015), https://unstats.un.org/unsd/gender/downloads/WorldsWomen2015_report.pdf (last accessed 15 July 2019), p. x.

[16]UN Department of Economic and Social Affairs, The World's Women: Trends and Statistics, 2015, ST/ESA/STAT/SER.K/20 (2015), https://unstats.un.org/unsd/gender/downloads/WorldsWomen2015_report.pdf (last accessed 15 July 2019), p. xiii.

[17]UN General Assembly Resolution "Transforming our world: the 2030 Agenda for Sustainable Development", A/RES/70/1 (25 September 2015), para. 20.

[18]UN General Assembly Resolution "Transforming our world: the 2030 Agenda for Sustainable Development", A/RES/70/1 (25 September 2015), preamble, 2nd para.

It is noteworthy that the UN member states emphasize the importance of gender equality together with the empowerment of women.[19] Indeed, the realization of the SDGs for women is inextricably linked with women's ability and opportunity to assert their fair share in the implementation of the 2030 Agenda. After all, it is their future, too, that is shaped through the SDG implementation processes. Therefore, women need to be agents of that change. The gender gaps between women and men as describes above are not simply the results of individual choices, but the results of gender-based power relations within all states of the world. They cause structural discrimination of women in the political, economic, and social fields, which can only be overcome by women's full and equal participation on all levels.

3 The Convention on the Elimination of All Forms of Discrimination Against Women

A Globally Binding Standard

Eradicating individual and structural discrimination of women in all areas—political, economic and social life—is the purpose of the Convention on the Elimination of All Forms of Discrimination against Women (CEDAW). It was adopted by the UN General Assembly in 1979, entered into force in 1981, and has been ratified by 189 states.[20] Until today, states, politicians, and scholars call into question the characterization of CEDAW as *the* global treaty on women's human rights because of the high number of reservations made by states upon ratification.[21] However, a closer analysis reveals that the content and purposes of the reservations vary greatly.[22] Many reservations have been removed or modified by now, reflecting changes in the understanding of the religious or cultural motivations (as well as of the reasons of domestic politics) that had prompted them upon ratification.[23] Moreover, many states have not made reservations to other UN human rights treaties insofar as they guarantee substantive human rights and gender equality. On the contrary, in the Vienna World Conference and the Beijing World Conference on

[19]UN General Assembly Resolution "Transforming our world: the 2030 Agenda for Sustainable Development", A/RES/70/1 (25 September 2015), para. 20.

[20]Ratification status as of 15 July 2019. Conspicuously missing are Iran, Somalia, Sudan, the US, and the Holy See.

[21]Tiefenbrun (2012), pp. 52–53. However, CRC has more reservations to substantive provisions, see Keller (2014), p. 311.

[22]For an analysis of the types of reservations see: Marsha A. Freeman, Reservations to CEDAW: An Analysis for UNICEF, UNICEF: New York December 2009, https://www.unicef.org/gender/files/Reservations_to_CEDAW-an_Analysis_for_UNICEF.pdf (last accessed 15 July 2019), pp. 6–7. On the types of "Islamic reservations" see Mayer (1998), pp. 26–45. On the permissibility of the reservations see: Lijnzaad (1994).

[23]Connors (2012), pp. 591–594.

Women, states have collectively proclaimed women's human rights as human rights, as being universal, inalienable, indivisible, and interrelated.[24] Thus, the politically motivated attempts to downplay the importance of CEDAW as legally guaranteeing the universal human rights for all women, do not hold up to closer legal scrutiny.

It is noteworthy that, despite the existence of a reservation, the CEDAW Committee discusses the substantive issues covered by it with the state party concerned during the monitoring procedure under the Convention. It even addresses recommendations on the issues to them. This approach is based on the Committee's understanding that reservations are permissible so as to give a state time to adapt its legal and factual situation to the Convention, but not to opt out of parts of it permanently.[25] Thus, the recommendations of the CEDAW Committee are intended to help the state achieve the aim of full realization of all Convention rights. This fact makes them useful for SDG implementation as well (see below at Sect. 4).

Addressing the Root Causes of Discrimination

CEDAW is unique among human rights treaties because it does not prohibit discrimination on the grounds of sex, but discrimination *against women*. This asymmetrical approach is due to the insight that there is a categorical difference between discrimination that women experience daily all over the world and discrimination against men: Discrimination against women is embedded in gendered societal power relations. They influence individual conduct, and more importantly, they also permeate structures, procedures, and institutions of the state as well as within society and the family. These power relations are upheld by gender stereotypes, which express societal expectations of women's (and men's) proper conduct and whose violations are sanctioned. They ensure a hierarchy between men and women and the domination of women by men. Thus, CEDAW reflects the understanding that discrimination against women is not based on sex—*i.e.* biological differences between women and men—but on gender—*i.e.* social constructions of what makes a woman and what makes a man. Its focus on gendered societal power relations is also the reason why CEDAW expressly obligates states to work towards overcoming gender stereotypes (Article 5).

[24] Vienna World Conference on Human Rights, Vienna Declaration and Programme of Action of 25 June 1993, A/CONF.157/23, para. 18 ("The human rights of women and of the girl-child are an inalienable, integral and indivisible part of universal human rights"); Fourth World Conference on Women, Beijing 4-15 September 1995, Final Declaration, para. 9, A/CONF./177/20/Rev.1 (1995) ("human rights of women and of the girl child as an inalienable, integral and indivisible part of all human rights and fundamental freedoms").

[25] By this approach the Committee sidesteps the legal issues of the consequences of reservations that are incompatible with the object and purpose of CEDAW, and the question of whether it is competent to decide on the compatibility of a reservation with the Convention.

By focusing on power relations CEDAW recognizes that women's full and equal enjoyment of all human rights requires a transformation of the distribution of power in the state, society, and the family. This understanding translates into a definition of discrimination that revolves around the exclusionary effect of discrimination (Article 1).[26] Consequently, CEDAW expressly prohibits not only legal, but also factual discrimination (Article 2(f)). In addition, the state's obligation to eliminate discrimination not only covers discrimination by state actors, but by private actors as well (Article 2(e)). Finally, Article 4 permits temporary special measures aimed at accelerating *de facto* equality between women and men. These pillars make for the transformative potential of CEDAW.

Articles 7 to 16 expound the states' obligations in further detail so as to ensure equal participation, empowerment, and full access to resources. The provisions encompass women's equal participation in political and public life, both domestically (Article 7) and internationally (Article 8), as well as women's equality in becoming a citizen (Article 9). Equal access to education (Article 10) is the basis for women's empowerment. Access to resources is guaranteed through the right to equality in the labour market (Article 11), in health care (Article 12), and in all areas of economic, social and cultural life (Article 13). For many of these rights, full and equal legal capacity is a necessary prerequisite, which is, therefore, enshrined in Article 15. Full access to resources also hinges upon women's capacity to hold, inherit or otherwise acquire, keep and manage property, also during and after marriage, and to engage in salaried work or other economic activities. Article 16 lays down these rights for women. It also ensures women's self-determination with respect to marriage, *viz.* entering into marriage with a freely chosen spouse, the equal rights and responsibilities as a spouse and as a parent. Thus, CEDAW requires abolishing norms and practices by which husbands can hold power over wives. Lastly, CEDAW safeguards women's self-determination over their bodies, lives and future by expressly enshrining the "right to decide freely and responsibly on the number and spacing of their children" and to "access to the information, education and means to enable them to exercise these rights" (Article 16(1)(e)). Women's reproductive rights are reinforced by the guarantee of equal access to health care services related to family planning (Article 12(1)). The content of many of these rights has been spelt out in more detail by the CEDAW Committee through its General Recommendations.[27] These authoritative interpretation guidelines help

---

[26]Discrimination is defined as "any *distinction, exclusion or restriction* made on the basis of sex which has the *effect or purpose* of *impairing or nullifying* the recognition, enjoyment or exercise by women, irrespective of their marital status, on a basis of equality of men and women, of human rights and fundamental freedoms in the political, economic, social, cultural, civil or any other field." (emphasis added).

[27]Committee on the Rights of Persons with Disabilities (CRPD) and the Committee on the Elimination of All Forms of Discrimination against Women (CEDAW), Joint statement "Guaranteeing sexual and reproductive health and rights for all women, in particular women with disabilities", 29 August 2018, INT/CEDAW/STA/8744 (2018).

states develop their implementation policies and direct the monitoring under CEDAW's state reporting procedure.

It is noteworthy that CEDAW considers women in rural areas an important cross-cutting issue deserving a specific provision (Article 14). This is an illustration of the Convention's approach to take into account the power relations in societies, as women in rural areas are often in particularly vulnerable situations due to the lack of public services and infrastructure, distance of state authorities, under-development, or poverty. The Committee has devoted a General Recommendation to the provision.[28] True to its power-sensitive approach, it also addressed issues of how to deal with the intersection of gender-based discrimination and other grounds of discrimination in a General Recommendation.[29]

What is conspicuously absent from the text of CEDAW is violence against women. When the Convention was drafted, violence against women was considered a question of social policy, not of human rights. It was only through the work of the CEDAW Committee, based on the findings from scholars and reports from women activists all over the world that this understanding changed. In 1993, the UN General Assembly recognized that "violence against women is a manifestation of historically unequal power relations between men and women, which have led to domination over and discrimination against women by men and to the prevention of the full advancement of women, and [...] violence against women is one of the crucial social mechanisms by which women are forced into a subordinate position compared with men."[30] This insight had led the CEDAW Committee, a year earlier, to adopt General Recommendation 19.[31] It characterizes violence against women as a form of gender-based discrimination because it targets women disproportionately or because they are women. In 2017, the CEDAW Committee refined its understanding in General Recommendation 35.[32] This document provides a comprehensive human rights based analysis of the forms and dimensions of violence against women, taking into account their diversity. It thus provides an excellent blueprint for monitoring states' compliance with their obligations under CEDAW to prevent, combat and redress violence against women effectively.

---

[28]CEDAW Committee, General recommendation No. 34 (2016) on the rights of rural women, CEDAW/C/GC/34, 7 March 2016.

[29]CEDAW Committee, General recommendation No. 28 on the core obligations of States parties under article 2 of the Convention on the Elimination of All Forms of Discrimination against Women, CEDAW/C/GC/28, 16 December 2010.

[30]UN General Assembly Declaration on the Elimination of Violence against Women, A/RES/48/104 (20 December 1993), 6th preamble paragraph.

[31]CEDAW Committee, General recommendation No. 19: Violence against women, INT/CEDAW/GEC/3731 (1992).

[32]CEDAW Committee, General recommendation No. 35 on gender based violence against women, updating general recommendation No. 19, CEDAW/C/GC/35, 27 July 2017.

A Blueprint for Comprehensive Action

This overview shows that the full and effective elimination of the structural discrimination of women in the political, economic, and social fields are at the core of CEDAW. The human rights guaranteed by the Convention give women a legal basis for their demands, thus contributing to women's empowerment. As described above, these objectives are identical to those of the SDGs. Consequently, SDG implementation must tackle, as a priority, the structural causes of gender inequality, *viz.* unequal participation in private and public decision-making, violence against women, unpaid care work, and limited control over assets and property.[33] In the words of the CEDAW Committee: "Realizing the full enjoyment of human rights by women is at the core of the 'transformative' impact of the Agenda 2030."[34] Using CEDAW and the insights gained through the monitoring process carried out by the CEDAW Committee should, therefore, guide the implementation of the SDGs.

Like all other UN human rights treaties, the implementation of CEDAW by the states parties is monitored by a committee of independent experts: the CEDAW Committee. It reviews the reports that every government must submit every four years after ratifying the Convention. The CEDAW Committee also receives information from non-governmental organizations and national human rights institutions, compares them with the state reports and discusses them with the government in interactive dialogues. This way, the CEDAW Committee has identified implementation gaps in which State action is necessary. The recommendations that the Committee addresses to each state party are public.[35] States are expected to ensure follow-up of the recommendations with the participation of all relevant domestic actors. In this vein, CEDAW has led to, or contributed to national and subnational actions plans—human rights action plans, gender equality action plans, or specific plans such as action plans to combat violence against women. During the next reporting cycle before the CEDAW Committee, the state party is held to account on whether and what actions it has taken and their effects. Through this iterative process over several decades, the CEDAW Committee has been able to identify core women's human rights issues for almost each of the countries in the world, and is continuously evaluating the effectiveness of measures taken. SDG implementation should make use of this wealth of information.

The monitoring process under CEDAW also made it easier for states, non-state actors and international organizations to understand the linkages between different

---

[33]UN Women, A Transformative Stand-alone Goal on Achieving Gender Equality, Women's Rights and Women's Empowerment: Imperatives and Key Components, New York 2013, http://www.unwomen.org/-/media/headquarters/attachments/sections/library/publications/2013/10/unwomen_post2015_positionpaper_english_final_web%20pdf.pdf?la=en&vs=1454 (last accessed 15 July 2019), pp. 2–3.

[34]CEDAW Committee, Submission to the Hugh Level Political Forum, 2016, INT/CEDAW/INF/8697 (2016), p. 4.

[35]They are available through the OHCHR Treaty Body Database at https://tbinternet.ohchr.org/_layouts/15/treatybodyexternal/TBSearch.aspx?Lang=en.

women's human rights issues and across sectors. For example, gender-based stereotypes cause gender-based discrimination in society leading to many women focusing (or having to focus) on care work at the household level and to low access to economic resources and decision-making. This, in turn can lead to boys being favoured to pursue an education, resulting in a repetition of the cycle in the next generation. In order to break this cycle, it is not sufficient to address one element only; each of them has to be seen in its interlinkages. If, for example, women's economic participation is encouraged through training programmes or micro-credits, it is also necessary to ensure that family law does not give the husband the right to decide on financial assets in the family and that the law ensures all women's rights to conclude contracts. In addition, the state has to develop ways to ensure that family care work is not the sole responsibility of women. Such interlinkages are reflected in the country-specific recommendations of the CEDAW Committee. This is a further reason for using CEDAW within the SDG implementation.

4 Using CEDAW for Promoting Gender-Sensitive SDG Implementation

According to the 2030 Agenda, the primary responsibility for follow-up and review lies with the states.[36] It envisages follow-up and review of the SDGs at the national, regional, and global level.[37] At the global level, this is assured through the High Level Political Forum (HLPF), composed of the UN member states, the major groups identified by Agenda 21, and other stakeholders with a standing invitation to participate as observers in the General Assembly.[38] Women constitute one of the major groups; they are represented by the Women's Major Group, a coalition of more than 600 organizations working to advance gender equality and women's human rights. The HLPF holds thematic discussions reviewing progress, e.g. by exchanging examples of good practice or specifying indicators, and they discuss the voluntary national reviews submitted by states. Regional follow-up and review varies, but in all regions the five regional UN Economic Commissions are involved. The design of follow-up and review at the national level forms is determined by each state, ideally after an inclusive participation process, with the support of the UN country teams, and that of other stakeholders, such as partner countries.

---

[36]UN General Assembly Resolution "Transforming our world: the 2030 Agenda for Sustainable Development," A/RES/70/1 (25 September 2015), para. 47.

[37]UN General Assembly Resolution "Transforming our world: the 2030 Agenda for Sustainable Development," A/RES/70/1 (25 September 2015), para. 47, and UN General Assembly Resolution "Follow-up and review of the 2030 Agenda for Sustainable Development at the global level," A/RES/70/299 (29 July 2016), paras. 2 and 10.

[38]UN General Assembly Resolution "Format and organizational aspects of the high-level political forum on sustainable development," A/RES/67/290 (23 August 2013), para. 14.

It is striking that, despite the strong commitment to gender mainstreaming in the 2030 Agenda,[39] the resolution on follow-up and review does contain any provision of how to ensure that this commitment is honoured. The participation of the Women's Major Group permits a gender perspective to be brought to the HLPF,[40] but there is no obligation of the states to consider, let alone accept, proposals in this respect. Similarly, in his progress reports to the HLPF, the UN Secretary-General applies a gender perspective.[41] Here again, it depends on the states whether they use this information in the thematic follow-up and review debates or in the debates on the voluntary national reviews. As show above (Sect. 3.3), the core concepts of CEDAW and the results of the international monitoring process under the Convention are a prime practical reason for using CEDAW to ensure a gender-sensitive implementation of the SDGs.

Advantage CEDAW: Legally Binding Force

In addition to the practical argument for having CEDAW guide states' SDG implementation, there is a strong legal argument that states are also legally bound to do so. The 2030 Agenda expressly provides for its implementation "consistent with the [...] obligations of states under international law",[42] which includes human rights conventions such as CEDAW. This reflects a fundamental rule of international law: States must comply with their human rights obligations in whatever action they take. Consequently, states must make sure they do not violate women's human rights when implementing the SDGs. The 2030 Agenda goes even beyond this minimum by affirming, in clear and explicit language, the object and purpose of the CEDAW Convention, when acknowledging the need for realizing women's human rights to achieve the SDGs (see above, at Sect. 2).[43] This congruence means that states must actively use the SDG implementation to realize women's human rights.

The legal force of CEDAW is of utmost practical relevance: Despite their high visibility and their present importance on the global level, the SDGs remain, in the

---

[39]UN General Assembly Resolution "Transforming our world: the 2030 Agenda for Sustainable Development," A/RES/70/1 (25 September 2015), paras. 20 and 74 (e).

[40]See, e.g., Women's Major Group, Paper for High Level Political Forum (2016). "Ensuring that no one is left behind": Listen to women for a change, 2016, https://wedo.org/wp-content/uploads/2016/07/10118WMG_HLPF_paper_2016_27April.pdf (last accessed 15 July 2019).

[41]See, e.g., UN Secretary-General, Special edition: progress towards the Sustainable Development Goals, E/2019/68 (2019).

[42]UN General Assembly Resolution "Transforming our world: the 2030 Agenda for Sustainable Development," A/RES/70/1 (25 September 2015), para. 18.

[43]UN General Assembly Resolution "Transforming our world: the 2030 Agenda for Sustainable Development," A/RES/70/1 (25 September 2015), para. 20, in particular; "The achievement of full human potential and of sustainable development is not possible if one half of humanity continues to be denied its *full human rights* and opportunities." (emphasis added).

end, voluntary political commitments.[44] As can be observed world-wide, translating the political commitment expressed at the global level into political action at the national level is not a given. When women's equal share in benefiting from the implementation of the SDGs in their respective country is at stake, gendered power relations will constitute powerful barriers, as they do in many political decisions. Through CEDAW, however, advocates for gender-justice in the sustainable development can bolster their demands through the legal obligation for all state actors to respect, protect and fulfill women's human rights, including by integrating these rights into the national legal systems, their political strategies, and by allocating sufficient budget for the implementation measures. Using CEDAW this way is particularly important for (state and international) actors in development cooperation, because it permits them to counter the criticism by powerful domestic actors that their call for women's rights means imposing "foreign values." Advocates for gender-justice in SDG implementation can point to CEDAW as the binding global standard to which the state has adhered by its own volition, and that, consequently, the state has to take serious the recommendations of the CEDAW Committee addressed to it.

Concluding Observations and Recommendations as Baselines and Targets

The concluding observations expressed by the CEDAW Committee at the end of a state reporting cycle provide a good overview of the main challenges that a state faces in the areas covered by CEDAW; and the CEDAW Committee's recommendations identify what needs to be done. The concluding observations are the outcome of a participatory analysis of the information provided by the state, civil society, national human rights institutions, and international organizations, as the CEDAW Committee holds (formal or informal) exchanges with all these actors. For this reason, they usually provide a description of the country situation that is well-founded in quantitative and qualitative data. Thus, they can be used as a baseline, on the basis of which future progress can be measured. This is what happens in the next reporting cycle before the Committee, but it can also be used for the purposes of SDG implementation.

The level of detail of recommendations varies: Some of them formulate a target to reach, others spell out in detail which law or policy has to be adapted in which way so as to eliminate discrimination against women. In most cases, the recommendations leave room for the state authorities to develop a solution that is tailored to the

[44]This follows from the non-binding character of UN General Assembly resolutions. Moreover, UN General Assembly Resolution "Transforming our world: the 2030 Agenda for Sustainable Development," A/RES/70/1 (25 September 2015), para. 74 (a), expressly affirms that follow-up and review will be country-led.

political, legal, and cultural specificities of the country. When the CEDAW Committee recommends a very specific measure to take, this may even be used as an indicator for the implementation of the human right concerned.

5 The Role of the CEDAW Committee in SDG Follow-Up and Review Processes

The CEDAW Committee shares its expertise with the High-Level Political Forum (HLPF) on a regular basis. In its contributions to the 2016 HLPF, the Committee reminded states that the prohibition of discrimination against women aims not only at formal equality (*de iure* equality), but at substantive equality (*de facto* equality). Furthermore, it reminds states of the intersectional discrimination that women experience worldwide, and as it is concerned that "the main gap between the Agenda 2030 and the CEDAW standards concerns sexual and reproductive health and rights (SDG Target 5.6), it called upon the states to ensure CEDAW-compliant SDG implementation in this respect.[45]

In its submission to the 2017 HLPF session,[46] the CEDAW Committee emphasizes that violence against women has not diminished, including in the most developed countries. On equal participation in public life, the Committee's analyses show that positive trends remain limited in a number of ways: Participation of women in decision-making positions in all sectors—parliament, government, public service, private sector, academia—peaks at levels well under parity. Even where women are represented, this representation does not always translate into actual political power. That means, for instance, that if women head government ministries, those will often be the least powerful governmental departments. On access to resources and economic empowerment, the CEDAW Committee acknowledges that women are gaining more access to the labour market, but it highlights that when women are employed, they are more likely to occupy jobs that are insecure, hazardous, poorly paid, and not covered by social protection. These considerations illustrate the need for qualitative indicators in addition to quantitative ones.

For its contribution to the 2018 HLPF,[47] the CEDAW Committee uses its general recommendations and results from the state reporting procedure so as to highlight its specific concerns with respect to the Goals discussed at that session: Goal 4 (education), Goal 13 (climate change), and Goal 16 (access to justice).

---

[45]CEDAW, Submission to the 2016 High-Level Political Forum (2016), https://tbinternet.ohchr.org/Treaties/CEDAW/Shared%20Documents/1_Global/INT_CEDAW_INF_8697_E.pdf (last accessed 15 July 2019), pp. 2, 3, and 4-5, respectively.

[46]CEDAW, Submission to the 2017 High-Level Political Forum (2017), http://www.ohchr.org/Documents/HRBodies/CEDAW/CEDAW_HLPF17_28.04.2017.pdf (last accessed 15 July 2019).

[47]CEDAW, Submission to the 2018 High-Level Political Forum (27 April 2018), https://tbinternet.ohchr.org/Treaties/CEDAW/Shared%20Documents/1_Global/INT_CEDAW_INF_8699_E.pdf (last accessed 15 July 2019).

The CEDAW Committee not only shares its expertise with the HLPF for its thematic reviews. It also brings the SDGs into its interactive dialogues with governments. In its recommendations to Belarus,[48] Burundi,[49] Micronesia,[50] and Antigua and Barbuda,[51] for example, the Committee made very specific suggestions on how the government could address SDG targets 5.1 on discrimination against women; target 5.2 on violence against women; target 4.5 on educational disparities; and targets 3.1 and 3.7 on maternal mortality and sexual and reproductive health. Other recommendations refer to target 5(c) on sound policies and enforceable legislation for the promotion of gender equality by spelling out necessary content of an envisaged national action plan; to target 5.5 on women's equal participation in political and public life; or to target 8.5 (full employment and decent work for women) through an employment policy that includes temporary special measures and to increase women's recruitment by employers and through food subsidies, maternity allowances, and loans for women in the informal sector.[52] In addition, the CEDAW Committee now formulates a standard recommendation on the national SDG implementation process to ensure a gender perspective,[53] and even adds more specific procedural recommendations, such as mainstreaming the monitoring of its own recommendations into the State party's realization of the SDGs.[54]

---

[48]CEDAW, Concluding Observations on the eighth periodic report of Belarus, CEDAW/CO/BLR/CO/8 (2016), paras. 23, 52.

[49]CEDAW, Concluding Observations on the combined fifth and sixth periodic reports of Burundi, CEDAW/C/BDI/CO/5-6 (2016), paras. 13, 29, 35, 39.

[50]CEDAW, Concluding Observations on the combined initial to third periodic reports of the Federated States of Micronesia, CEDAW/C/FSM/CO/1-3 (2016), paras. 13, 33, 37.

[51]CEDAW, Concluding Observations on the combined fourth to seventh periodic report of Antigua and Barbuda, CEDAW/C/ATG/CO/4-7 (2019), paras. 14, 28, 39.

[52]CEDAW, Concluding Observations on the seventh periodic report of Angola, CEDAW/C/AGO/CO/7 (2019), paras. 16, 32, and 38, respectively.

[53]CEDAW, Concluding Observations on the combined fifth and sixth periodic reports of Burundi, CEDAW/C/BDI/CO/5-6 (2016), para. 55; CEDAW; Concluding Observations on the combined fourth to seventh periodic report of Antigua and Barbuda, CEDAW/C/ATG/CO/4-7 (2019), para. 7; CEDAW, Concluding observations on the fourth periodic report of Botswana, CEDAW/C/BWA/CO/4 (2019), para. 7; CEDAW, Concluding Observations on the ninth periodic report of Colombia, CEDAW/C/COL/CO/9 (2019), para. 7; CEDAW; Concluding Observations on the eighth periodic report of Ethiopia, CEDAW/C/ETH/CO/8 (2019), para. 7; CEDAW, Concluding Observations on the fourth periodic report of Serbia, CEDAW/C/SRB/CO/4 (2019), para. 7; CEDAW, Concluding Observations on the eighth periodic report of the United Kingdom of Great Britain and Northern Ireland, CEDAW/C/GBR/CO/8 (2019), para. 7. For the initial version see CEDAW (2016), Concluding Observations on the eighth periodic report of Belarus, CEDAW/CO/BLR/CO/8 (2016), para. 52.

[54]CEDAW, Concluding Observations on the fourth periodic report of Serbia, CEDAW/C/SRB/CO/4 (2019), para. 10.

Moreover, the CEDAW Committee also shares its expertise to help develop global SDG indicators for which methodologies and data are still missing.[55] For this purpose, it is engaging with UN Women and the Office of the High Commissioner for Human Rights. In an early position paper, the three entities presented suggestions for indicators and identified which existing UN bodies or entities could serve as monitoring agencies.[56] Nevertheless, the latest version of the United Nations Minimum Set of Gender Indicators, adopted by the United Nations Statistical Commission in 2018, barely reflects a human rights approach.[57] In contrast, the expert group that developed indicators on Goals 10, 13, and 16 for the 2019 HLPF, refers, in passing, to states' obligations under CEDAW and lists some general recommendations of the CEDAW Committee among its resources.[58] However, the focus of all these debates remains on quantitative data. Even if the process of developing human rights indicators for the various human rights treaties is still ongoing at the UN, it is clear that quantitative measuring has to be accompanied by qualitative data if the measuring exercise is to produce not only a picture of the factual situation, but also to permit evaluating whether there is room for bringing about better results in the future. With respect to CEDAW, there are proposals for qualitative indicators building on the CEDAW Committee's interpretation of the Convention, e.g. legislative indicators.[59] These could also be used in monitoring SDG implementation.

[55]The UN Statistical Commission calls these indicators "Tier III" indicators (differentiating them from Tier I and Tier II indicators for which methodologies and/or data do exist), https://unstats.un.org/sdgs/iaeg-sdgs/tier-classification/ (last accessed 15 July 2019).

[56]UN Women, Monitoring Gender Equality and the Empowerment of Women and Girls in the 2030 Agenda for Sustainable Development, New York September 2015, http://www.unwomen.org/-/media/headquarters/attachments/sections/library/publications/2015/indicatorpaper-en-final.pdf?la=en&vs=212 (last accessed on 15 July 2019).

[57]United Nations Statistics Division, The United Nations Minimum Set of Gender Indicators, 30 April 2019, https://genderstats.un.org/files/Minimum%20Set%20indicators%202018.11.1%20web.pdf (last accessed 15 July 2019). Note that under the heading "Human rights of women and girl children", the Minimum Set only comprises indicators related to violence against women.

[58]Expert Group Meeting, Tackling global challenges to equality and inclusion through the gender-responsive implementation of the 2030 Agenda for Sustainable Development: Spotlight on SDGs 10, 13 and 16, Vienna, 27–28 February 2019, Annex I, p. 34, published by UN Climate Change, UNODC, and UN Women, 2019, https://sustainabledevelopment.un.org/content/documents/23808EGMViennaFin.pdf. (last accessed 15 July 2019), p. 14, 27, and 50.

[59]International Knowledge Network of Women in Politics (joint project of the International Institute for Democracy and Electoral Assistance (IDEA), the Inter-Parliamentary Union (IPU), the United Nations Development Programme (UNDP), and the United Nations Entity for Gender Equality and the Empowerment of Women (UN Women), CEDAW Legislative Compliance Indicators, without date, http://iknowpolitics.org/sites/default/files/indicators_chapter.pdf (last accessed on 15 July 2019).

6 The Role of National Human Rights Institutions in SDG Implementation and Follow-Up

CEDAW will only guide the implementation of the SDGs if domestic actors and the international community push for this approach. An institutional actor whose mandate is to hold the state to its human rights obligations and to help it meet these obligations are National Human Rights Institutions (NHRIs).

NHRIs are independent public bodies with a broad mandate to protect and promote *all* human rights in their respective countries. They exist in more than 110 states worldwide. Their mandate and the criteria for independence are outlined in the Paris Principles, adopted by the UN General Assembly.[60] Each NHRI's compliance with the Paris Principles is regularly reviewed through an accreditation procedure carried out by the global network of NHRIs, the Global Alliance of National Human Rights Institutions (GANHRI), under the auspices of the UN Office of the High Commissioner for Human Rights.[61] NHRIs advise their governments and parliaments, they monitor their state's compliance with its human rights obligations, a lot of them can carry out investigations and bring cases to courts. Many NHRIs have an ombuds function or a complaint mechanism, through which people can submit their individual cases of alleged human rights violations. NHRIs are expected to engage with international human rights mechanisms so as to build a bridge between the international and the domestic levels and thus "bring human rights home." Thus, NHRIs can collect valuable information, they are experts in understanding international human rights and their implementation, and they hold governments to account. All these powers can be used with respect to CEDAW-compliant SDG implementation.[62]

Immediately after the adoption of the 2030 Agenda, NHRIs adopted the Mérida Declaration at their tri-annual global conference, and encouraged each other to contributing to the follow-up and review of the SDGs in a human rights compliant way.[63] In particular, they recommended action on the country level and at the global

---

[60]UN General Assembly Resolution "National Institutions for the promotion and protection of human rights," A/RES/48/134 (20 December 1993), Annex.

[61]For details of this procedure, see: GANHRI, A Practical Guide to the Work of the Sub-Committee on Accreditation, Geneva 2018, https://nhri.ohchr.org/EN/AboutUs/GANHRIAccreditation/Documents/GANHRI%20Manual_online(1).pdf (last accessed 15 July 2019).

[62]See also: Stephen L.B. Jensen, Allison Corkery, Kate Donald, Realizing Rights through the Sustainable Development Goals: The Role Of National Human Rights Institutions, Danish Institute for Human Rights, June 2015, pp. 4–6, https://www.humanrights.dk/files/media/dokumenter/udgivelser/research/nhri_briefingpaper_may2015.pdf (last accessed 15 July 2019).

[63]GANHRI, Mérida Declaration—The Role of National Human Rights Institutions in implementing the 2030 Agenda for Sustainable Development, 10 October 2015, https://nhri.ohchr.org/EN/ICC/InternationalConference/12IC/Background%20Information/Merida%20Declaration%20FINAL.pdf (last accessed 15 July 2019), para. 17.

level, as well as cooperation between NHRIs, especially through their global and four regional networks to strengthen the capacities of all NHRIs in this respect. The UN Human Rights Council expressly welcomed this approach of NHRIs, and encouraged NHRIs to implement it.[64]

At country level, NHRIs can provide advice on a human rights-based approach to implementation and measurement of the SDGs. This includes assessing the impact of laws, policies, programmes, national development plans, administrative practices and budgets from the perspective of the realization of all human rights for all. On a practical level, this means that NHRIs can promote *national targets, benchmarks and indicators* that take into account the respective country's human rights challenges and obligations when governments develop or revise national SDG implementation plans. These targets, benchmarks and indicators should be based on the human rights treaties ratified by the state, such as CEDAW, and the concluding observations of the respective treaty body, such as the CEDAW Committee. Germany's NHRI, the German Institute for Human Rights, produced a synopsis of the SDGs and the recommendations addressed to the state by the UN treaty bodies.[65] As the revised national sustainability strategy merely discussed violence against women, but did not measure it,[66] the German Institute for Human Rights stressed the need to include an indicator that measures the implementation of the Council of Europe Convention on Violence against Women, the so-called Istanbul Convention, which the CEDAW Committee had also emphasized in its recommendations to Germany. The next revision of the German Sustainable Development Strategy saw the announcement of a programme of action on the implementation of that Convention, while still falling short of including an indicator in this respect.[67]

---

[64]UN Human Rights Council, Resolution 39/17 of 28 September 2018 (A/HRC/RES/39/17).

[65]German Institute for Human Rights, Sind die SDGs für Deutschland relevant, Annex, September 2015, https://www.institut-fuer-menschenrechte.de/fileadmin/user_upload/Publikationen/Weitere_Publikationen/Sind_die_SDGs_fuer_Deutschland_relevant.pdf; English summary in: German Institute for Human Rights, Are the SDGs relevant for Germany?, October 2015, https://www.institut-fuer-menschenrechte.de/fileadmin/user_upload/Publikationen/Weitere_Publikationen/Are_the_SDGs_relevant_for_Germany.pdf (both last accessed 15 July 2019).

[66]German Federal Government, German Sustainable Development Strategy. New Edition 2016 (11 January 2017), https://www.bundesregierung.de/resource/blob/998220/455740/7d1716e5d5576bec62c9d16ca908e80e/2017-06-20-langfassung-n-en-data.pdf?download=1 (last accessed 15 July 2019), p. 96.

[67]Bundesregierung, Deutsche Nachhaltigkeitsstrategie. Neuauflage 2018 (German Federal Government, German Sustainable Development Strategy. Revised Edition 2018) (7 November 2018), pp. 33, 42–45, English version not yet available), https://www.bundesregierung.de/resource/blob/975292/1559082/a9795692a667605f652981aa9b6cab51/deutsche-nachhaltigkeitsstrategie-aktualisierung-2018-download-bpa-data.pdf?download=1 (last accessed 15 July 2019).

When reviewing policy options and their potential effects, NHRIs can conduct *human rights impact assessments*, for example to help the government decide on a policy to spur economic growth that will best promote decent work and fair pay for women. In Germany, the NHRI, together with civil society organizations, successfully advocated for including human rights in the national definition of sustainability, which is used to conduct impact assessments on new laws and regulations.[68]

National Human Rights Institutions can also advise their respective governments on how to establish *national SDG monitoring mechanisms* that are truly participatory and that ensure accountability of the State to its people. This can include a transparent process for civil society to share data and their own analyses, when the national SDG implementation plan is reviewed and adjusted.

At the global level, when their country presents its voluntary national review at the High-Level Political Forum, National Human Rights Institutions can share their own evaluation, and they can encourage civil society to prepare alternative reports, as the Women's Major Group recommends.[69] In addition, NHRIs can provide valuable information on the general obstacles to transparent and participatory SDG implementation. For this reason, GANHRI annually published a report on the status of civil society space in the states under voluntary national review, based on the information provided by the NHRIs concerned.[70] NHRIs have more to contribute, including on women's human rights, as they committed to increase work in this field

[68]Bundesregierung, Deutsche Nachhaltigkeitsstrategie. Neuauflage 2018 (German Federal Government, German Sustainable Development Strategy. Revised Edition 2018) (7 November 2018), pp. 33, 42–45, English version not yet available), https://www.bundesregierung.de/resource/blob/975292/1559082/a9795692a667605f652981aa9b6cab51/deutsche-nachhaltigkeitsstrategie-aktualisierung-2018-download-bpa-data.pdf?download=1 (last accessed 15 July 2019), p. 51.

[69]Women's Major Group, Recommendations from the Women's Major Group on engaging with the Voluntary National Review Process, 2017, http://www.womenmajorgroup.org/wp-content/uploads/2017/05/WMG_Info_Note_Final_2017.pdf (last accessed 15 July 2019).

[70]GANHRI, Protecting and enlarging the space for public debates and participation of all civil society actors for the implementation of the SDGs and human rights. Background Paper (for HLPF 2016), 12 July 2016, http://nhri.ohchr.org/EN/News/Documents/GANHRI-BackgroundPaper-HLPF16-ShrinkingSpace_publ.com.pdf; GANHRI, Protecting and enlarging the space for public debates and participation of all civil society actor for the implementation of the SDGs and human rights. 2nd Background Paper (for HLPF 2017), July 2017, https://nhri.ohchr.org/EN/News/Documents/20170709_GANHRI-BackgroundPaper%20-HLPF17-ShrinkingSpace_final.pdf#search=shrinking%20space; GANHRI, Space for Civil Society Participation in SDG Implementation. 3rd Background Paper (for HLPF 2018), July 2018 (all last accessed on 15 July 2019).

in their Amman Declaration of 2012.[71] The work of NHRIs in the area of women's human rights is reflected in their parallel reports to the CEDAW Committee in the state reporting procedure. Moreover, GANHRI has collected information from NHRIs on specific, CEDAW- and SDG-related themes, such as economic participation of women,[72] the human rights of women in rural areas,[73] and violence against women.[74]

The UN General Assembly encouraged National Human Rights Institutions explicitly to participate and contribute to the deliberations on the 2030 Agenda.[75] However, NHRI participation is severely limited as NHRIs and their global and regional networks do not constitute a major group or other stakeholder (see above at Sect. 4). The UN bodies are still to heed the call of the General Assembly to extend NHRIs' participation rights.[76] An invitation by the President of the General Assembly would make the wealth of information from NHRIs available to the international community for a better, CEDAW-compliant and gender-responsive implementation of the 2030 Agenda.

---

[71] ICC—International Coordinating Committee of National Human Rights Institutions (today: Global Alliance of National Human Rights Institutions (GANHRI), Amman Declaration and Program of Action, 7 November 2012, https://nhri.ohchr.org/EN/ICC/InternationalConference/11IC/Background%20Information/Amman%20PoA%20FINAL%20-%20EN.pdf (last accessed 15 July 2019).

[72] Asia-Pacific Forum (APF) & Global Alliance of National Human Rights Institutions (GANHRI), The Role of National Human Rights Institutions in Promoting and Protecting Women's Economic Participation in the Changing World of Work, Concept note presented at the 61st session of the Commission on the Status of Women 2017, https://nhri.ohchr.org/EN/Themes/WomenRights/Commission%20on%20the%20Status%20of%20Women/CSW_61_The%20role%20of%20NHRIs%20in%20Promoting%20and%20Protecting%20Women%E2%80%99s%20Economic%20Participation%20in%20the%20Changing%20World%20of%20Work.pdf (last accessed 15 July 2019).

[73] Global Alliance of National Human Rights Institutions (GANHRI) & Asia Pacific Forum (APF), The Role of National Human Rights Institutions in promoting gender equality and the empowerment of women and girls living in rural areas, Report presented at the 62nd session of the Commission on the Status of Women 2018, https://nhri.ohchr.org/EN/Themes/WomenRights/Commission%20on%20the%20Status%20of%20Women/GANHRI-APF%20report%20on%20the%20Role%20of%20NHRIs%20in%20promoting%20gender%20equality%20and%20the%20empowerment%20of%20women%20and%20girls%20in%20rural%20areas.pdf (last accessed 15 July 2019).

[74] Global Alliance of National Human Rights Institutions (GANHRI), Preventing and Eliminating All Forms of Violence against Women and Girls: The Role of National Human Rights Institutions A contribution to the review and priority themes of CSW63, Geneva 2019, https://nhri.ohchr.org/EN/Themes/WomenRights/Commission%20on%20the%20Status%20of%20Women%2062/DIMR_GANHRI%20CSW%20Report_final%20BF.pdf (last accessed 15 July 2019).

[75] UN General Assembly Resolution "National institutions for the promotion and protection of human rights", A/RES//70/163 (17 December 2015), paras. 15.

[76] UN General Assembly Resolution "National institutions for the promotion and protection of human rights", A/RES//70/163 (17 December 2015), paras. 16.

7 Conclusion

The transformative potential of the Sustainable Development Goals hinges on women's equal participation, their empowerment, equal access to resources, and freedom from gender-based violence. Gender-based discrimination and exclusion of women is deeply entrenched in all societies and state structures worldwide. Through the work of the CEDAW Committee, CEDAW provides an invaluable resource for identifying the core challenges in every state, and the Committee's concluding observations and recommendations provide a baseline for future evaluation and specific targets. The CEDAW Committee, as well as National Human Rights Institutions, have been using this qualitative information so as to achieve a CEDAW-compliant and gender-responsive implementation of the SDGs. CEDAW and the 2030 Agenda are mutually reinforcing: CEDAW brings legally binding force, detailed obligations, and its own monitoring mechanism, which it applies for the SDGs as well. The SDGs bring and renew the political commitment and the international cooperation necessary to ensure that the future development of the world is sustainable and does not leave behind half of humanity.

References

Connors J (2012) Article 28 CEDAW. In: Freeman MA, Chinkin C, Rudolf B (eds) The UN Convention on the elimination of all forms of discrimination against women: a commentary. Oxford University Press, Oxford, pp 565–595

Keller LM (2014) The impact of states parties' reservations to the Convention on the elimination of all forms of discrimination against women. Mich State Law Rev 2014(2):309–326

Lijnzaad L (1994) Reservations to United Nations human rights treaties: ratify and ruin? Brill/ Nijhoff, Leiden

Mayer AE (1998) Islamic reservations to human rights conventions. A critical assessment. RIMO. Recht van de Islam/Human Rights Islam 15:26–45

Sen G (2019) Gender equality and women's empowerment: feminist mobilization for the SDGs. Glob Policy 10(Supp. 1):28–37

Tiefenbrun SW (2012) Women's international and comparative human rights law. Carolina Academic Press, Durham

Beate Rudolf is Director of the German Institute for Human Rights (GIHR), the National Human Rights Institution of Germany (A-status accredited under the United Nations Paris Principles). From 2016 to 2019, she also chaired the Global Alliance of National Human Rights Institutions (GANHRI). Prior to her appointment to the GIHR, she was a junior professor for public law and equality law at the faculty of law of Freie Universität Berlin and director of the research project "Public International Law Standards for Governance in Weak and Failing States" within the Research Center "Governance in Areas of Limited Statehood". Her research focuses on human rights and legal principles on state structures under public international law, European law and German constitutional law. She was Vice-President of the European Women Lawyers Association (EWLA) and is a member of the ILA Committee on Feminism in International Law as well as an advisory member of the Committee on Public International and European Law of the German Women Lawyers' Association (djb). She is a co-editor of the international CEDAW Commentary published by Oxford University Press and was awarded the Margherita-von-Brentano-Award of Freie Universität Berlin for her outstanding academic and socio-political contribution to human rights, in particular women's human rights.

SDGs, Gender Equality and Women's Empowerment: What Prospects for Delivery?

Josephine A. Odera and Judy Mulusa

Contents

Abstract This paper is conceptualized within the framework of gender equality and women's empowerment and proceeds from the premise that the developmental and political goal of reducing gender inequalities remains largely unmet. The 17 SDGs with goal 5 as a stand-alone on gender equality and more than half of the 17 other goals have integrated gender dimensions with measurable indicators. The pioneers for women's rights, over a century ago, focused on the labor market. Today, exclusion and discrimination in the labor market indicate that the struggle against these inequalities remains valid. Drawing on the discourse on poverty, the paper notes that this is a fundamental issue for the SDGs but that the feminization of poverty puts more women at risk. An institutional perspective to gender equality and women's empowerment beckons, if the SDGs are to deliver on this cross-cutting agenda. The paper considers the 2015 review of progress since the adoption of the Beijing Platform of Action (BPFA) undertaken at the same time as the transition from the MDGs to the SDGs and poses the question, can the SDGs deliver on gender

J. A. Odera (✉)
Africa Centre for Transformative and Inclusive Leadership, Nairobi, Kenya

J. Mulusa
Department of Business and Entrepreneurship, Prince Georges Community College, Largo, MD, USA

M. Kaltenborn et al. (eds.), *Sustainable Development Goals and Human Rights*, Interdisciplinary Studies in Human Rights 5,
https://doi.org/10.1007/978-3-030-30469-0_6

equality and women's empowerment. The review noted that while there has been progress at the normative level, overall progress, has been unacceptably slow, with stagnation and even regression in some contexts. Change towards gender equality has not been deep enough, nor has it been irreversible (United Nation, Res 69/313, Addis Ababa Action Agenda for the third international development conference on financing and development, 2015). The paper analyses the potential for the promise of the SDGs to make change irreversible against the background that while economic prospects appear to have risen, gender disparities have persisted and in some instances, widened, despite the common knowledge that closing the gender gap portends even greater economic growth. The paper argues for a transformative approach that can address deep structure, capacities, mindset and organizing.

1 From MDGs to SDGs

The 2030 Agenda for sustainable development came into effect on 1st of January 2016. It is an improvement of the previous agenda, the Millennium Development Goals (MDGs) which were supposed to have been met by 2015. The MDGs focused on poverty and its alleviation, a digression from the broader conception of development which had earlier focused on enlarging the productive capacity of economies to make possible improved living standards. It was composed of 8 MDGs and 21 targets all focused on ending extreme poverty. They were credited for reviving the interest in development issues in areas of poverty, education and health and strengthening the willingness to put more resources into aid. However, the MDGs were criticized for applying a narrowly defined view of development with very few indicators, giving them a minimalist message, which only served to distract states from previous state commitments. The MDGs employed aggregate targets which did nothing for within country inequalities but focused resources on particular goals.[1]

The goals directed a western perspective of development and poverty eradication on developing countries with limited financial commitment from developed countries. They completely left out issues such as; peace, security, human rights, democracy and good governance. MDG 3 which called for promotion of gender equality and empowerment of women had one target: elimination of gender disparities in primary, secondary and tertiary education by 2015. Critics assert that this target left out all issues needed to achieve gender equality and women's empowerment including eliminating violence on women, recognizing the burden of unpaid care work that disproportionately falls on women, women's limited access to assets and resources,

[1]For example in health the focus was on disease specific goals (vertical programs) and this led to separate strategic plans, monitoring mechanisms, funding streams and implementation effort. The result was that there was achievement of health indicators while major deficiencies in the health systems persisted.

protecting women's sexual and reproductive health rights and ending harmful traditional practices such as early and forced marriage and female genital mutilation.

The 2030 Agenda goes beyond poverty and covers a set of issues across three dimensions of development: economic, social and environmental. It is universal as it takes a holistic approach to addressing the challenges of sustainable development and it applies to all countries rather than to developing countries only. The sustainable development goals (SDGs) address key challenges such as combatting climate change and achieving gender equality not only through a stand-alone goal but in a cross-cutting manner therefore capturing the interconnectedness of development concerns. The SDGs were a result of an inclusive process with a high level of ownership from government, unlike the MDGs which were based on the millennium declaration and developed by a number of 'UN insiders'. The open nature of formulating the SDGs permitted civil society organizations, UN agencies and private corporations to engage at various points in the drafting phase. The SDG universality acknowledges that progress on sustainable development 'must draw a deeper theoretical base' and the new goals embrace a form of institutional cosmopolitanism. Universal targets in the SDGs are grounded in the idea of global public goods; the environment, health and economic growth annulling the distinction between the developing and developed countries.

Compared to the MDGs, the agenda of the SDGs is broader, overall and with respect to gender and potentially more transformative. It addresses many more aspects of women's lives and therefore represents a considerable advance on the MDGs. The 2030 Agenda prioritizes gender equality as a stand-alone goal (SDG 5) and a cross cutting issue. It has its foundation in the Beijing Declaration and Platform for Action (BPFA) and the Program of Action of the International Conference on Population and Development. Up till 2015 the BPFA was the world's most powerful framework for gender equality policies and practices. It recognized the role of discriminatory laws, beliefs and practices in driving gender inequalities such as violence, early marriages and unpaid care work. The platform highlighted the need to forge a universal agenda for change and integrated social norms throughout the 12 critical areas with targets specifically addressing discriminatory institutions.[2] It envisioned gender equality in all dimensions of life. A review of the BPFA in 2015 showed significant gains as governments increasingly removed discriminatory laws, girl's enrollment in secondary schools increased, women's participation in the labor market increased and birth control increased. Nevertheless, global progress has been slow, uneven and limited. A number of challenges in implementation have hindered the achievement of BPFA's vision including persistent discriminatory social norms and gender stereotypes, low levels of participation and leadership of women in decision making at all levels, chronic underinvestment in gender equality globally and absence of strong accountability mechanisms. This notwithstanding, opportunity exists to take advantage of the demographic dividend and make changes to lay the foundation of equality and sustainability.

[2]Such as early marriages, unequal inheritance and land rights and gender-based violence.

Coming from the MDGs where the commitment to gender equality was limited to targets on gender parity in education and maternal mortality, the SDGs stand-alone goal on gender is more comprehensive and potentially a transformative commitment for women's rights. Target 5.1 to eliminate all forms of discrimination takes up Beijing's call to address discriminatory social norms, attitudes and stereotypes as well as discriminatory laws. The SDGs incorporate gender specific targets in other goals, for example: to eliminate gender disparity in education (SDG 4.5), ensure women's access to adequate sanitation (SDG 6.2), equal pay for work of equal value, (SDG 8.5), and safe and affordable transport for women (SDG 11.2). The SDGs also include 'means of implementation' as a goal (SDG 17) and as targets for each goal, admitting the need to change policies and institutions if transformative change is to take place. Inclusion of these issues indicates that the SDGs are more holistic and more aligned to gender equality than the MDG's. The most important question for the SDGs is one of effectiveness. Does the agenda and the theory of change embedded in the SDGs hold promise and are its elements transformational for women and girls? In what follows we review some of the structural elements in the SDGs that may be of current or future concern in achieving transformative changes in gender equality and women's empowerment.

2 Extent of Gender Inclusion

Women's empowerment and gender equality have a catalytic effect in the achievement of human development,[3] however, the 2015 review of the Beijing platform still concluded that no country had fully achieved gender equality and empowerment of women and girls. Agenda 2030 confirms the centrality of women's equality and women's and girl's empowerment, several goals in the SDGs include at least some mention of gender under their associated targets. This is true of the goal on poverty (goal 1), nutrition (goal 2), education (goal 4), water and sanitation (goal 6), employment (goal 8) and cities (goal 11). Target 1.b calls on countries to '. . .create sound policy frameworks as the national, regional and international levels, based on pro-poor and gender sensitive development strategies. . .'. The inclusion of this targets and other gender specific targets reflects the notion that when it comes to certain issues such as health, education, poverty and nutrition there are gender specific experiences that require attention. However, a number of the SDGs make no reference to gender; goal 7 (ensure access to affordable, reliable, sustainable and modern energy for all) makes no mention of the gender dimensions of energy when it

[3]Desai (2010), Onaran (2015), Nayana and Das (2017), Sinha et al. (2017) and Marone (2016). The papers look at the effect of gender equality and women's empowerment policy in wage led development strategy, land inheritance rights, child nutrition, gross development product (GDP), demographic dividend and poverty reduction.

has been acknowledged that energy policy is not gender neutral[4]; goal 9 (build resilient infrastructure and promote inclusive and sustainable industrialization) does not take account of the fact that poor infrastructure exacerbates the gender gap and makes no mention of the need to adopt gender sensitive indicators[5] to build a base for successful gender approaches in infrastructure; goal 12 (ensure sustainable consumption and production patterns) is silent on gender; the adaptation target in goal 13 (climate change) is not linked to gender, yet gender disparities increase vulnerability of women and girls to climatic risk and adaptation initiatives that do not take gender into account may unintentionally replicate gender inequality[6]; and goal 16[7] (peace and inclusive societies) lacks gender specific language. To effectively work towards achievement of gender equality and women's empowerment gender must be integrated across all SDGs and gender considerations must be included in all sustainable development work.

3 Accountability and Monitoring

Global initiatives designed to benefit the daily lives of people risk not being implemented appropriately if at all, unless they are monitored and accountable. The selection of appropriate results and indicators is essential to strengthen accountability for implementation of the SDGs and particularly to ensure gender equality and women's empowerment. Gender indicators were first developed during the Fourth World Conference on Women in Beijing in 1995. Since then attempts to transpose broad questions and aspirations into gender indicators remains a daunting task. There is an emphasis on quantitative targets at the expense of context, history and culture—for instance there is no other measure of poverty other than the money metric (USD 1.25 per day) which fails to capture many of the hardships that constitute poverty, for example child labor, unpaid women labor, undernourishment, illiteracy, exposure to violence and lack of access to water, shelter and sanitation, factors that affect women disproportionately. The same argument may be applied to the measure of unpaid care work. In 1995 the Beijing platform highlighted the

[4]Gender dimensions in energy can be captured by examining links between energy policies and women's welfare, productivity and empowerment, see Sharma et al. (2016), Half et al. (2014) and O'Dell et al. (2014).

[5]Such as: estimate of time savings per trip for women, improved frequency of trips to health clinics and change in cooking practices.

[6]Climate change adaptation is inherently context specific and often based on changing behavior implying that effective gender responsive adaptation responds to unique gender dynamics of the target area, Wolf (2011).

[7]Explanations put forward to account for the exclusion of gender in this goal include; the relatively strong governance orientation of gender equality in SDG 5, political tension surrounding SDG 16 prior to its adoption and compression of the goal necessitated by the merging of what had originally been two separate goals, see Goetz and Jenkins (2016a).

importance of tackling the unequal distribution of paid and unpaid care work. Focus on unpaid care work within the framework of the SDGs echoes Beijing's call.

Razavi notes that the wording, 'recognize and value unpaid care work' draws attention to counting unpaid work in statistical terms, which has its value in terms of bringing the issue of unpaid care work to the limelight, but the action on this target should go a couple of steps further to promote shared responsibilities between men and women.[8] Chopra notes that "economic empowerment is not just about women's participation in the labor market but participation in a way which takes into account their unpaid work responsibilities which is decent, well paid and which means flexible working hours and a choice of locations."[9]

Secondly, a number of indicators do not appear to capture the full intention of the targets in the SDGs whereas there are some targets that do not capture the full intention of the goal. Goal 5 has six targets; one of them, target 5.2, is to 'eliminate all forms of violence against all women and girls in the public and private spheres, including trafficking and sexual and other types of exploitations'. This target is measured by two indicators 5.2.1 '... proportion of ever partnered women and girls aged 15 years and older subjected to physical, sexual or psychological violence by a current or former intimate partner in the last 12 months by form of violence and age group' and 5.2.2 '...proportion of women and girls aged 15 years and older subjected to sexual violence by persons other than an intimate partner in the last 12 months, by age group and place of occurrence'. The broad goal has been subdivided into a more limited target and the indicators used, measure only a small part of the issues in the target. The two indicators measure frequency with reference to age, type of violence and type of relationship. Issues such as trafficking and other forms of exploitation are not captured by the indicators. These indicators make no mention of the experience of violence, the trajectory of the relationships, the cultural meanings of gender, marriages and sexuality as well as the social context of the violence amongst other factors.[10]

Establishing the transformative nature of the SDGs begs the question are the targets specific, measurable, attainable and time bound? Pogge and Sengupta note that in some instances the SDGs are not specific enough and in instances where the SDGs include specific means of implementation they fail to specify who is responsible for each goal and or target.[11] They mention that the target on education parity which is an improvement over the MDGs[12] is lacking in terms of specific crucial indicators pertinent for girls: distance to school; safety of schools; travel to school;

[8]Razavi (2016). Measuring promotion of shared responsibilities within households and family creates another challenge.

[9]Chopra (2015).

[10]Merry (2016).

[11]Pogge and Sengupta (2015).

[12]It has shifted focus from enrollment in primary schools to include secondary education and early childhood learning while expanding the focus from enrollment to completion.

early marriages and pregnancies[13]; target 5.5-on representation of women at national or local government level lacks an aspirational percentage and metrics to measure women's levels of influence in power and this is the same concern raised about the targets under the energy goal (goal 7). Target 5.1-end all forms of discrimination against all women and girls everywhere, fails to specify the efforts required to achieve this and it also fails to hold a particular agent accountable while target 2.1—end hunger and ensure access by all people in particular the poor people in vulnerable situations, fails to assign effort to specific actors/agents. Koehler criticizes the long time frame associated with some of the targets/indicators[14]; for instance the accomplishment of the targets on education parity and reduction in maternal mortality were pushed to the year 2030 while the global maternal mortality target's ratios-less than 70 per 100,000 live births, are less ambitious than the targets that high income countries have set for themselves-16 per 100,000 live births.[15]

To emphasize the problem of gender indicators and or targets it is worth noting that during the drafting phase of Agenda 2030, there was a proposal by gender equality lobbyists to include a target on stronger women's rights organizations in the SDG framework.[16] This target did not make it into the final SDG framework because of the complexity of measuring the size and strength of women's movements, a number of measures that have been proposed in literature have proved to be inadequate.[17] Over and above this, out of the 230 unique global SDG indicators 53 explicitly reference women, girls, gender or sex, including the 14 under SDG 5 however, UN Women finds that there are no internationally established methodologies or standards for 23 out of the 53 gender related indicators.

Lastly lack of data further compounds the inadequacies of the gender indicators, it makes it harder to accurately identify, analyze and monitor the separate needs and vulnerabilities of women, girls, men and boys and develop effective evidence-based policies and solutions. A study by Data 2x,[18] has reviewed the availability of sex disaggregated data and found that while 80% of countries regularly produce sex disaggregated data statistics on mortality, labor force participation and education and training, less than one third of countries disaggregate statistics on informal

[13]Early marriages and pregnancies were listed by the BPFA as major causes of girls lower educational performance and higher dropout rates causing the gap in school completion rates for girls.

[14]Koehler (2016).

[15]United Nations (2015), p. 38.

[16]Goetz and Jenkins (2016b).

[17]For example: number of women active in women's organizations or number of women's organizations in a country, capacity of civil society actors to accumulate political and social resources to support gender equality and proportion of managers of civil society organizations and media professionals that are women Htun and Weldon (2012), Mazur et al. (2012) and UN Women (2013).

[18]An initiative started by Hilary Clinton to improve the quality of gender data and its analysis to drive better policy making http://www.data2x.org/.

employment, entrepreneurship, violence against women and unpaid work.[19] Lack of disaggregated data masks the differences among various groups making it challenging to identify the needs of marginalized populations and it results in a review mechanism that is based entirely on the interpretation of government officials. Given the weak and undemocratic nature of many global economic or political governance bodies this becomes a big challenge for implementation of the SDGs in general. Investment in data offers a critical impetus to revive Beijing's commitment to address key data gaps by strengthening national capacities.

4 Method of Implementation

One of the limitations in implementing the MDGs stemmed from the lack of an 'opaque system' to monitor and report progress. There were no institutional mechanisms through which beneficiaries could engage in shaping or challenging decisions at the domestic level in a meaningful way.[20] Situating SDGs in international law and politics by considering their legalization, is important in determining the successful implementation of the SDGs. Legalization, a particular form of institutionalization shows the degree to which imposition of international legal constraints on governments can be applied. It refers to a set of characteristics defined as a continuum based on three dimensions; obligation, precision and delegation.[21] Scholars like Kenneth Abbot and Duncan Snidal distinguished harder from softer international law[22] based on the three dimensions; obligations- harder law will have a higher degree of legalization and softer law will have weak or no legal obligations; precision—institutions may be written in more or less detailed and precise language with harder law having a high degree while softer law comprises vague, general or abstract wording and; delegation—in harder law interpretation or enforcement is left to independent third parties[23] while with softer law interpretation and enforcement is kept within parties allowing for political maneuvering.

SDGs are seen to be at the soft end of the continuum; there are no hard obligations, legal rules and commitments that impose a particular type of binding obligation[24]; precision which narrows the scope of interpretation is present in some targets[25] but is generally low as most targets do not unambiguously define certain

---

[19]Buvinic et al. (2014).

[20]Nayyar (2012) and McArthur (2013).

[21]Abbott et al. (2000), Karlsson-Vinkhuyzen and Vihma (2009) and Goldstein et al. (2000).

[22]Hard legalization is where all three properties are maximized or are 'high' and soft legalization involves different combinations of the 3 attributes.

[23]Like an international tribunals, courts and arbitrators.

[24]Breach of legal obligations creates a 'legal responsibility' which is lacking in SDGs.

[25]See target 3.4 'reduce by one third premature mortality from non-communicable diseases'.

conduct but rather they state vague and aspirational outcomes[26]; delegation of authority to designated third parties to implement agreements, is lacking as the SDG framework appears to embrace political bargaining between parties who may accept or reject proposals. The fact that SDGs are a statement of aspirations: a voluntary agreement rather than a binding treaty presents an opportunity for states to adopt a more ambitious agenda but creates a drawback in so far as states may be more tempted to skirt their commitments. It is noted that the commitments and precision in SDGs fall short of those made in the Beijing Platform. The SDGs aim broadly for equal access to justice while the BPFA specifically commits governments to ensuring access to free or low cost equal services designed to reach women living in poverty; secondly, SDGs call for promotion of peaceful and inclusive society while BPFA called for reduction in excessive military expenditure and controlling availability of arms. Moreover, target 5.4 'recognize and value unpaid care work... as nationally appropriate', target 5.1 'end all forms of discrimination' and target 5.5 'ensure women's full and effective participation' are open to many interpretations since there are no clear definitions of unpaid care work, discrimination and effective participation. On a more general note lack of precision is observed under the goal on ending poverty, there is no commitment to a collectively maintained universal social protection floor, but merely the call for 'nationally appropriate social protection systems and measures'. In conclusion the soft nature of the SDG laws is emphasized by the language in the resolution adopting the 169 targets-'[they] are defined as aspirational and global with each government setting its own national targets guided by the global level of ambition but taking into account national circumstances...[27]'. This wording begs the question which targets will be considered more important when countries develop their own targets?

5 Macroeconomic Framework

The type of economic thinking dominating current global macroeconomic policy has been described as neoliberalism. It places competition at the center of human relations and sees a diminished role of government. It is characterized by austerity measures, global integration and increased production. Feminists highlight that these policies are gender neutral. They do not take into account economic and social rights, distributive outcomes or gender equality. Indeed, they typically focus on a narrow set of goals, such as raising economic growth rates or reducing inflation to extremely low levels. These measures are not necessarily linked to more fundamental objectives such as human development, well-being and the enjoyment of rights. Feminist economists have demonstrated a number of points which cast doubt on the assumed link between gender equality and economic growth. For example, with

[26]Only 29% of targets in SDGs are well defined (Hazelton 2015).

[27]Agenda 2030 paragraph 55.

financial globalization the costs of austerity measures, privatization of government assets and services, cuts in public expenditure, welfare policies and social protection have historically been and continue to be borne by women as women's unpaid care work acts as a stabilizer of economic financial crisis.[28] Women are effectively assuming the costs of current failure to regulate markets in the form of unemployment, low wages, and job insecurity. Unregulated global financial speculation has also led to the rise in prices that destabilize household budgets, and impact women as caregivers and as small farmers. BPFA openly challenged austerity programs and the impact of these macroeconomic policies on women. The Platform acknowledged that the neo liberal "trade not aid"-model of development was and is failing the majority of the world's women.

Agenda 2030 contains a number of targets that aim at guarding against global inequality and problems of future financial crises: goal 17 has a target to enhance macroeconomic stability and another to respect each country's policy space; target 10.6 makes mention of enhanced representation; goal 8 promotes economic growth with a hope for increased productivity, technological change and resource efficiency; goal 12 mentions sustainable production and consumption patterns; and goal 10 emphasizes reduction of inequality within countries, that is, attain growth for the bottom 40% of the population at a rate that is higher than the national average. However, Ponte and Enriquez assert that Agenda 2030 does not explicitly recognize the link between women's human rights, gender equality and the global economic governance policies.[29] Analyzing the aforementioned SDGs shows why this is the case. Goal 10, on inequality relies on the old model of industrial growth; with emphasis on increasing levels of extraction, production and consumption. It calls for at least a 7% annual GDP growth in least developed countries and higher levels of economic productivity across the board. Target 10.11 concentrates on the bottom 40% of the population making it a pro poor growth model. The target wants to reduce poverty by 'ratcheting' the poor up while leaving the wealth and power of the top 1% intact. The target makes no mention of the link between poverty and exploitation and monopolization of resources by the few wealthy countries. This inequality goal does not ask for a specific level of growth in incomes of the poorest but simply any growth at all above the national average. Goal 8 is devoted to economic growth, specifically export oriented growth, in keeping with neoliberal models. The emphasis on the economic growth approach fails to respond to macroeconomic and structural drivers as well as environmental limitations of growth. Increasing incomes does not automatically translate into gender equality or better well-being for females,[30] relying solely on growth is no panacea. Whether growth contributes to gender equality depends on growth patterns, that is, which sectors drive economic growth and if they are capable of generating decent employment for women. It also depends on the role of the state in redistributing growth and the

[28]Ponte and Enriquez (2016) and Antonopoulous (2014).

[29]Ponte and Enriquez (2016).

[30]See Razavi (1997).

presence of structures that curtail women's ability to take advantage of the growth. Relying on economic growth is a supply side story, it is based on the premise that individual countries are ultimately responsible for their development provided their policy space is respected (target 17.15). Goal 17.10 calls for more trade liberalization but is silent on the unfair trade regime of the World Trade Organization (WTO) and the many bilateral trade and investment agreements.[31] In many countries, trade liberalization has led to a reduction in government revenues, with implications for policies to support gender equality. Trade liberalization increases gender inequality in the labor market[32] as women's low wages render them the workforce of choice for the export market. Emphasis on export markets leads policy makers and producers to see wages as a cost rather than a source of potential demand, leading to strategies to reduce wages. Trade agreements between countries also often have provisions that limit the policies that individual governments can adopt, such as measures that aim to promote and support domestic productive activities and investment. Such agreements frequently have different consequences for women's and men's employment. Looking back at the effects of the policies in the 1980s and the 1990s by the World Bank and the International Monetary Fund (IMF) makes this more evident, as the increase in exports depressed prices and led to deterioration in the terms of trade. Recent trade agreements like the Trans Pacific Partnership (TPP) grant stronger corporate control of the world's resources and trade flows. This presents enormous risks for realization of human rights, ecological social and gender justice. Lastly the SDGs are silent on financial speculation and the need for greater regulation of financial markets. Goal 17.13 talks of the need to 'enhance global macroeconomic stability' through policy coordination but gives no specific targets while no mention is made of tax evasion and avoidance which drains developing countries.

6 Role of the State and the Private Sector

SDGs explicitly call for involvement of a wide range of actors including the private sector. Progress on many of the 17 goals will depend on private sector contributions[33]-goal 12 on responsible consumption and production urges Transnational Corporations (TNCs) to adopt sustainable practices and integrate sustainability information while goal 17 emphasizes partnerships. The Addis Ababa Action Agenda (AAAA), the outcome document for finance and development, underlines the importance of the TNC's in achieving SDGs. Needless to say, the overall impact

[31] Hickel (2015).

[32] See Farha et al. (2017); United Nations Inter-Agency Network on Women and Gender Equality.

[33] TNCs account for as much as 25% of global GDP (Patchell and Hayter 2013), the private sector accounts for 90% of jobs in less developed countries, foreign direct investment is outpacing development assistance 4 to 1.

of TNCs is multifaceted and complex and varies depending on a number of factors.[34] Cees Van Beers states that competition from TNCs creates a pervasive incentive structure that encourages states to go easy on regulations as well as economic and civil rights in the interest of fostering a suitable business environment and compliant labor force.[35] From this perspective discrimination against women is imperative and is a stepping stone towards more competitive status. Tied to the problem of TNCs is the issue of International Financial Flows (IFFs). IFFs also have a disproportionate gender impact, draining critical resources that could otherwise be allocated to advancement of women's human rights. IFFs from Africa result in estimated losses of over $ 500 billion per year with 65% due to commercial tax evasion and tax avoidance including transfer of income to tax havens. Unlike the SDGs the BPFA takes a stand on the IFFs stating that the government should analyze and adjust macroeconomic policies including taxation and external debt policy from a gender perspective to promote a more equitable distribution of productive assets. Current legal and political frameworks in many developing countries allow TNCs to benefit from tax evasion. Be that as it may, the 2030 Agenda lacks an internationally binding instrument to regulate activities of TNCs in international human rights law particularly in the global south. Relying on the TNCs may circumvent one of the SDG's central goals.

The SDGs do not clearly acknowledge the role of the states in delivering SDGs particularly in human rights and gender equality. Goal 17.16 which reads '... enhance the global partnership for sustainable development, complemented by multi-stakeholder partnerships...' may imply that states have a primary role, but the document does not state this explicitly. 'Partnerships' implies an increased role for the private sector a factor which is questionable based on the principles of equality. Public private partnerships (PPPs) have been promoted at the national level and by the United Nations development system as the best way to advance investments in areas of special relevance for women and human rights particularly in social infrastructure and social services. The perspective that PPPs can narrow the gender gap is questionable. Critics have argued that PPP projects do not deliver outcomes and the problems are heightened in countries with weaker monitoring and regulation mechanisms by the state.[36] Though the influence of PPPs emphasizes efficiency and effectiveness, the profit motives of the private sector threatens realization of human rights and would worsen gender equality. PPPs have changed how interventions are designed and managed, focusing more on showing value for money through quantifiable indicators. This is evident in health care, where existing PPPs do not contribute to alleviating fiscal pressure but rather exacerbate the fiscal

[34]Including local capacity of production and innovation, host government policies, investor motives the investment model and firm characteristics.

[35]Van Beers (2000).

[36]See Romero (2015).

unsustainability and often fail to deliver services needed by women especially those living in poverty.[37]

In tandem with the role of the private sector vis a vis the state, is the political will. A number of factors raise concern over the extent of the political will. During the preparation phase of the SDGs countries from Africa and Arab groups contested references to long agreed commitments on women's rights like the BPFA and the international conference on population and development. A record 143 countries guaranteed equality between men and women in their constitutions by 2014 but another 52 have not yet taken this step (UN Women, ESAR). With regard to specific gender targets in the SDGs it is noted that some countries strongly opposed the inclusion of sexual and reproductive health rights in the SDG framework, to the extent that the issue of sexual rights is still missing in the SDGs. Weak political commitment at the preparation phase of the SDGs was also observed when a number of governments rejected proposals for stronger financial regulation that would tap the wealth of the top 1% and curb illicit flow of finances from developing countries[38] and when countries in the global North refused to accept the creation of an inter-governmental tax UN body on tax matters.

7 Financing of SDGs

The review of the BPFA after 20 years showed that underinvestment in gender equality and women empowerment contributed to slow and uneven progress in all the 12 critical areas of concern. An overview of donor support shows low levels of financing-in 2012–2013, USD 4 billion targeted gender equality as a principal objective, representing 5% of all OECD-DAC aid, while women's economic empowerment received 2% of aid. By comparison, USD 22 billion or 25% targeted gender equality as a secondary objective. This adds up to only 30% of aid being responsive to women's needs and interests. Interestingly though, aid to gender equality has grown at a faster rate than aid overall. While total aid grew at a rate of 1% annually aid to gender equality grew by 4% annually. This figure was even higher in fragile states with an annual growth rate of 10%.[39] Despite this trend, investments in gender equality are vastly insufficient as the 2030 agenda is estimated to cost between USD 2–3 trillion every year for 15 years, this represents 4% of world's GDP yet developed countries have committed to provide only 0.7% of GDP in aid.

[37]See Gideon and Porter (2014).

[38]The final draft of the SDGs acknowledges the need for international financial institutions (IFIs) to continue to respect the policy space of each country consistent with IFIs organizational mandates which are very much set by the developed countries anyway. By adding this the SDGs missed an opportunity to transform the international financial institutions.

[39]OECD 2015 States of Fragility 2015: Meeting post 2015 Ambitions, OECD, Paris.

Aid in support of gender equality is concentrated in social sectors of health and education with significant underinvestment in gender equality in economic and productive sectors. In 2013 only 2%(ibid) of aid to the economic and productive sectors targeted gender equality as a principal objective. The transport and energy sectors receive the bulk of bilateral aid to climate change but only a very small proportion of this targets gender equality. In 2014 climate ODA that also supports the achievement of gender equality was USD 8 billion, accounting for 31% of bilateral ODA.[40] Only 3% of this, had gender equality as a principal objective. Ensuring climate financing instruments are responsive to women's needs and priorities remains a major concern in the SDGs. Currently only 20–24% of jobs in renewable energy sector are held by women, yet projections show that green jobs could lead to millions of new employment opportunities in the next few years. An analysis of the Nationally Determined Contributions (NDC) submitted by parties to the United Nations Framework Convention on Climate Change (UNFCC) shows that only 40% included any reference to gender and women.

The gap between policy commitments and financial commitments is also pronounced in the area of peace and security and sexual and reproductive health. In 2012–2013 only 2% of aid to peace and security in fragile states targeted gender equality as a principal objective. The same applies to aid to ending violence against women and girls. Funding for women's civil society organizations and institutions constitutes a very small percentage of total aid to gender equality. In fragile states women's equality organizations and institutions received just 1% of total aid targeting gender equality,[41] yet women's organizations are at the forefront of peacebuilding and state building efforts.

8 What Will a Transformative Approach Entail?

Hillenbrand et al. define a gender transformative approach as one that aims to move beyond individual self-improvement among women toward transforming the power dynamics and structures that serve to reinforce gendered equities.[42] Such an approach goes beyond the 'symptoms' of gender inequality to address the 'social norms, attitudes, behaviors and social systems that underlie them'. Rottach et al. ascribe gender transformative change to a set of issues including; promotion of the position of women, challenging resource distribution and allocation of duties between men and women and addressing the power relationships between men

[40]OECD 2015 Making climate finance work for women: Overview of bilateral ODA to gender and climate change, OECD Paris.

[41]OECD Aid in support of gender equality and women's empowerment. http://www.oecd.org/dac/stats/gender-related-aid-data.htm Accessed on 10/27/2017.

[42]Heillenbrand et al. (2015).

and women in the community.[43] The Interagency Gender Working Group notes that gender transformative approaches 'actively strive to examine, question and change rigid gender norms and imbalance of power'.[44] These definitions confirm Beijing's message that social norms play a critical role in accelerating progress toward gender equality. The challenge of transformative strategies is targeting the structural foundations of gender inequality and ensuring participation while, recognizing the diversity of girls and women's experiences to ensure the most marginalized are included in decision making even in the political process. Based on lessons learned to be transformative action is required in the five priority areas discussed below.

Strengthening Accountability Mechanisms

Establishment of a rigorous follow up and review mechanism that is transparent, state led, incorporates peer review and allows for participation by the civil society is key. Linking the SDGs and their targets to the Convention on Elimination of Discrimination Against Women (CEDAW) and the implementation of the BDPFA enables a feminist accountability framework where member states must report their progress and can be challenged by civil society. The Commission on the Status of Women (CSW) also has a critical role to oversee implementation of the 2030 Agenda.

Secondly accountability banks on clear indicators which will require complimenting the international targets/indicators with contextually relevant benchmarks in individual states. Gender sensitive indicators must be contextualized to multiple dimensions of disempowerment encountered by the female population. They should be ambitious and not resort to what is easy to measure and reflect need rather than availability of data. To come up with a list of indicators that is narrow enough to be usable, applicable and achievable for every country while remaining separate from any political influences requires balancing a wide range of competing priorities and technical limitations. Use of sex disaggregated indicators will allow tracking of the degree to which women have benefitted (or not) from whatever progress is made. A number of global accountability mechanisms and initiatives have been established to drive progress on data and accountability when it comes to issues affecting women.[45] Use of creative and progressive indicators to make targets more specific and incorporate a rights-based approach into the language of the indicators/targets is imperative.

---

[43]Rottach et al. (2009).

[44]Interagency Gender Working Group in Heillenbrand et al. (2015).

[45]These include the High Level Political Forum, Universal Periodic Review, The Independent Accountability Panel, Equal Measures 2030, Data 2X, Making Every Woman and Girl Count, Countdown to 2030 and the UN System Wide action Plan of Gender Equality.

Securing national ownership of gender related SDGs and targets in the early stages of implementation is an impetus to accountability. Governments voluntarily adopted the SDGs and they bear their overall responsibility. Localization—the process of defining, implementing and monitoring strategies at the local level for achieving global, national and subnational SDG targets is crucial. Coupled with deep legal and legislative changes to protect women's rights, governments should ensure that the capacity of national statistical organizations become more transparent, inclusive and capable of mainstreaming gender accountability and producing disaggregated data to guide interventions and hold governments accountable. Ownership of the SDGs calls for multi-channeled awareness raising campaigns at the national and local level, establishing and sustaining substantive dialogues among multiple stake holder bodies and supporting efforts of women machineries and CSOs to play leadership roles in the implementation of Agenda 2030.

Use of Soft Law to Establish a Consensus

Although soft law has not been acknowledged as law due to the lack of binding character, it has significant and extensive effects on establishment and development of international law. Soft law presents a flexible environment for consensus free from parliamentary approval or long years of application as is the case with customary law.[46] Increased compliance with soft law will require more specific rather than vague agreements, focus on regulating state actors rather than non-state actors, forming links with hard law, norms on absenteeism rather than those requiring actions and use of financial incentives.[47]

Transforming the Economy to Achieve Gender Equality

As part of the commitment to gender equality, macroeconomic policymaking should incorporate distributive impact analyses that allows for macroeconomic policies to be evaluated in terms of their effects on women and men. Specifically, there is a need to analyze the effects of public spending, tax policy and monetary policy on gender equality and where gender inequalities are uncovered, governments must take steps to correct them. Other policy areas with macro-level impacts, such as the negotiation of trade agreements, also need to be subjected to greater scrutiny. Trade policy must have a good understanding of where women are in the economy while the trade negotiators must be provided with information on sensitive sectors where trade liberalization should be expedited, delayed or exempted with a view to enhancing

[46] Kaltenborn and Kuhn (2017).

[47] Shelton (1997).

or protecting female employment or female owned enterprises. Less developed countries have traditionally been more cautious about including non-trade concerns in the trade agreements, they fear that these considerations may become a financial burden or may result in trade barriers. For these countries a transformative approach will require a paradigm shift to ensure gender equality needs are included in the core text of trade agreements, to increase political commitment. A number of examples are seen in current trade agreements: the Cotonou Agreement[48] states that '...countries should respect international conventions regarding women's rights and gender equality...'; the North American Agreement on Labor Cooperation, a side accord that accompanies the North American Free Trade Area (NAFTA) reads '... promote cooperative activities regarding among others (a) equality of women and men in the work place...'[49] and the US-Central America Free Trade Area embedded gender considerations within their capacity development mechanisms—the Labor Cooperation and Capacity Building mechanisms which sets out gender equality as a cooperation and capacity building priority.[50] Addressing trade imbalances is now even more urgent given the gigantic trade agreements being hatched across the world's largest economies. These agreements grant stronger corporate control over the world's resources and trade flows and this presents an enormous risk for the realization of human rights, ecological, social and gender justice.

To support global sustainable development, macroeconomic policies will have to target more closely real variables, such as the level of output and employment. Indeed, this may require shifting emphasis from intermediate targets such as low inflation, external balance, and low fiscal deficits. But moving away from intermediary goals does not mean that macroeconomic stability should be abandoned. On the contrary, a primary goal of policy is to ensure stable financial and monetary conditions in support of productive development. Thus, a critical task of macroeconomic policies, necessitates the reorientation of economic dynamics towards equality-oriented production, consumption and reproduction patterns with attention to transformative redistribution of resources. This requires extensive regulation of financial markets, market structures and competition and on redistributing mechanisms and labor markets. Macroeconomic policies to support sustainable development must give greater weight to domestic demand, policies that can be pursued by all countries simultaneously, without counter-productive wage and tax competition. In addition, central banks can use their regulatory power to channel credit to uses that support the realization of rights and promote gender equality. Policies and regulations can also encourage credit to be extended to improve housing, stimulate job-creating investments or promote access to loans for self-employed workers in informal activities, many of whom are women. Greater emphasis should be placed on developing redistributive economic policies and social policies that incentivize an

[48]http://ec.europa.eu/development/icenter/repository/agr01_en.pdf.

[49]http://www.dol.gov/ilab/regs/naalc/naalc.htm.

[50]www.ustr.gov/sites/default/files/uploads/agreements/cafta/asset_upload_file320_3936.pdf.

equitable distribution of household labor for example those that increase government revenue and accountability rather than cutting important basic services.[51]

Public investment spending in social infrastructure[52] would generate public employment in labor intensive social services and could increase female labor force participation rate via socializing the invisible and unpaid care work. An analysis done by feminist economist at the UK Women's Budget Group[53] shows that investing 2% of GDP in the care industry would increase women's employment rates between 3.3 and 8.2%. In the US according to this analysis, such an investment would create nearly 13 million new jobs, much more than investing 2% of GDP in construction sector, which would create around 7.5 million jobs. Some 67% of the new jobs created by investment in the care sector would go to women, compared to 35% of the new jobs from investment in the construction sector. These jobs need to be made more attractive for all, by improving pay and working conditions in these industries. A reorientation toward high skilled decent service sector jobs should be promoted instead of reliance on low pay service jobs with weaker labor unions. However, as much as public investment will reduce the amount of unpaid care work there is need to redistribute the remaining work equally between men and women. This requires action from governments, businesses, trade unions and women's organization to mobilize resources and change cultures.

Increasing Investments in Gender Equality

There is need for transformative financing that increases investment and resources to meet new commitments. Funding of women's rights organizations is critical and will require earmarking of resources. The Addis Ababa Action Plan[54] outlines transformative policy and financing actions to accelerate implementation of existing commitments in the Beijing Declaration and Platform of Action and to meet new commitments on gender equality and women's empowerment. It emphasizes the importance of progressive budgeting and or gender responsive budgeting and tax policies which have enormous scope to reduce gender inequalities. Removing gender biases in taxation systems could help level the playing field. Mobilizing domestic financial resources through increased domestic borrowing has been a trend in emerging market economies. Within the context of a prudent medium-term fiscal framework and a public debt management strategy this may be a useful source of

[51]Floro and Willoughby (2016).

[52]Including child care, education and elder care as well as water and electricity.

[53]Analysis done by feminist economists at the UK Women's Budget Group.

[54]Please see "Policy Hub – Transformative financing and goals for gender equality and women's and girls' empowerment – Addis Ababa and New York 2015", European Parliament, Directorate General for Internal Policies, Policy Department C: Citizens' Rights and Constitutional Affairs, Women's Rights & Gender Equality, July 2015. http://www.europarl.europa.eu/RegData/etudes/IDAN/2015/519237/IPOL_IDA(2015)519237_EN.pdf.

additional revenue. Such borrowing has a catalytic impact on domestic financial sector development but for developing countries accumulation of debt and the risk of experiencing debt distress is a concern.

OECD admits that for every US dollar which comes to developing countries as ODA three US dollars leave these countries as illicit financial flows. Hence reducing outflow due to illicit financial flows, debt service repayments and maintenance of foreign reserves in developed countries will mobilize domestic resources for southern countries. SDG 17, the global partnership goal, creates space for individual countries but shields the world's most powerful agents; the rich countries, international organizations and trans-national corporations (TNCs) from any concrete responsibilities in achieving the SDGs. Given their wealth and resources, they should take the lead in providing resources and implementing systematic institutional reform to address the root cause of poverty. These reforms would include changing the rules that encourage illicit financial outflows from developing countries or force poor countries to pay debt with interest on loans accumulated by illegitimate and unaccountable rulers. If the SDGs hold these affluent agencies accountable for their role in making the sustainable development work, then the concept of partnership and universalism in the SDGs would be more meaningful. National Assessment Plans (NAPs) should be designed to encompass SDGs while respecting human rights. For TNCs this means adopting a positive approach toward equal payment whilst ensuring TNC's activities are non-exploitative. Aligning corporate and SDG indicators is imperative. SDGs indicators should be set so that companies can set sustainability commitments related to their core business and report against them in line with the SDGs. Several TNCs already report against targets covered by SDGs such as environmental impact and women's empowerment. The NAPs should have remedy schemes for those affected by human rights violations.

Engendering Climate Change with a View to Attaining Gender Equality and Sustainable Development

It is important to remember that women are not only vulnerable to climate change, but they are effective actors or agents of change. This is recognized in the first UNFCC Gender Action Plan, finalized in November of 2017, that integrates gender into all aspects of climate policy. Areas identified as critical blocks in response to climate change are; mitigation, adaptation, technological transfer, financing, monitoring and reporting. Mitigation and adaptation should effectively address gender specific impacts of climate change in the areas of food security, agriculture and fisheries, water, biodiversity, health, human rights, peace and security. Secondly technological development related to climate change should take into account women's specific priorities, needs and roles and make use of their knowledge and expertise. Women's involvement in the development of new technologies can ensure that they are user friendly, affordable, effective and sustainable. Financing must be flexible enough to reflect women's priorities and needs. Active participation of women in development of funding criteria and allocation of resources for climate change initiatives is critical while gender analysis of all budget lines and financial instruments is needed. The recent integration of gender considerations into key multilateral climate finance mechanisms[55] are steps in the right direction. Countries must make use of the climate finance tools/resources[56] available for ensuring gender responsive implementation of climate change projects and they must encourage a paradigm shift in climate finance thinking to ensure gender equality and women's empowerment dimensions are mainstreamed within all climate finance governance structures. The last pillar, monitoring and reporting requires countries to track the implementation of climate change policies. This is a good opportunity to track the effectiveness of gender mainstreaming initiatives and gender impacts of climate policies at all levels. Gender considerations and women's issues, needs and contributions should be integrated across the planning and execution cycle of climate change policies and projects. Monitoring should also focus on ensuring that, climate change and other goals in SDGs must work toward a common purpose. For example, at present individual goals on energy access and tackling climate change could contradict each other—massive expansion of fossil fuel for example, would satisfy one goal and undermine the other. These potential conflicts can be mediated by strong integrative targets.[57]

Ensuring Full and Equal Participation of Women in Decision Making

A comprehensive approach is needed to increase women's participation in power and decision making. Such an approach would encompass temporary special measures to achieve gender balance in decision making bodies and capacity building and training initiatives to support women's political participation at local and national levels. This calls for use of gender quotas as transitional mechanisms within electoral systems that are non-conducive to equal participation[58] as well as constitutional and legal reform. In addition, governments will need to address institutional contexts of decision making to create more women friendly institutions and organizational

[55]Green Climate Fund (GCF) and Clean Investment Funds (CIF).

[56]UNDP Gender Responsive National Communication Toolkit, the Guide to Gender Mainstreaming in UNDP Supported GEF Finance Projects, the Capacity Building Package on Gender Mainstreaming in Mitigation and Technology Development and Transfer Interventions and the IUCN Climate Change Gender Action Plans (ccGAPs) (UNDP 2017).

[57]For example, decrease carbon intensity by increasing the share of renewable energy to 30% and increase energy intensity by 2.4% per year or scaling down biomass facilities to match the size of forest residues.

[58]Quotas address institutional barriers and demand action from institutional actors and power holders.

structures. In industrialized countries this has been accomplished through offering subsidized or work-based childcare, maternity and paternity leave, emergency leave for care givers and introducing flexi time. In tandem addressing masculine stereotypes of power, discriminatory party practices and threats of violence against women in electoral context is an important step. This will require engaging boys and men in supporting girls and women leaders as well as holding individuals who obstruct women's abilities accountable. An increased role for women in decision making also banks on use of appropriate indicators to measure women's participation, with a focus on bringing the importance of women's local level participation to the fore. Measuring progress qualitatively and quantitatively captures the complex and changing realities of women's lives.

9 Conclusion

Governments exude confidence about the impending positive impact of Agenda 2030 in relation to gender equality and women's empowerment, nevertheless the textual analysis in this paper points the picture that the SDGs may fail to be as transformative as envisaged despite some clear positives. The influence of SDGs may be most significant in their ability to change the idea of development from a purely economic approach to one that includes a host of issues. The SDGs constitute an improvement with regard to gender equality and women's empowerment however, there is no room for complacency as we move to the implementation stage. Given that gender equality is a cross cutting development issue unless it is addressed in a multidimensional way it will not become a reality and not one of the SDGs will be achieved without including women and girls as equal partners.

References

Abbott KW, Keohane RO, Moravick A, Slaughter AM, Snidal D (2000) The concept of legalization. Int Organ 54:17–35

Antonopoulous R (2014) Gender perspectives and the gender impact of the global economic crisis. Routledge, New York

Buvinic M, Nichols R, Koolwal G (2014) Mapping gender data maps. Data 2x report, United Nations Foundation, Washington DC, http://www.fsnnetwork.org/sites/default/files/Data2X_MappingGenderDataGaps_FullReport_0.pdf. Accessed 22 Nov 2017

Chopra D (2015) Balancing paid work and unpaid care work to achieve women's economic empowerment, IDS policy brief 83. IDS, Brighton

Desai M (2010) Hope in hard times: women's empowerment and human development. Human Development Research Paper 14. United Nations Development Program, Human Development Office New York

European Parliament. Challenges and achievements in the implementation of the millenium development goals for women and girls from a European union perspective. Brussels

Committee on Women's Rights and Gender Equality. http://www.europarl.europa.eu/RegData/etudes/etudes/join/2014/493049/IPOL-FEMM_ET(2014)493049_EN.pdf. Accessed 14 Nov 2017

Farha F, Zhaohua L, Mohammad I (2017) Trade liberalization and gender inequality in emerging economies – from the perspective of sustainable development goals. Asian Econ Financ Rev 7 (11):1075–1092

Floro M, Willoughby J (2016) Feminist economics and the analysis of the global economy: the challenge that awaits us. Fletcher Forum World Aff 40(2):15–27

Goetz A, Jenkins R (2016a) Feminist activism and the politics of reform: when and why do states respond to demands for gender-equality policies? UNRISD working paper, no. 2016–13. United Nations Research Institute for Social Development (UNRISD), Geneva

Goetz A, Jenkins R (2016b) Gender security and governance: the case of sustainable development goal 16. Gend Dev 24(1):127–137

Gideon J, Porter F (2014) Unpacking women's health in public private partnerships: a return to instrumentalism in development policy and practice. Wider Working Paper No. 2 2014/009 Helsinki United Nations University (UNU Wider). https://www.wider.unu.edu/sites/default/files/wp2014–009.pdf. Accessed 29 Oct 2017

Goldstein J, Kahler M, Keohane R, Slaughter A (2000) Introduction: legalization and world politics. Int Organ 54(3):385–399

Half A, Sovacool B, Rozhon J (2014) Energy poverty: global challenges and local solutions. Oxford University Press

Hazelton A (2015) Africa: Science and Post 2015 SciDev.Net. http://allafrica.com/stories/201502190939.html. Accessed 27 Dec 2017

Heillenbrand E, Karim N, Mohanraj P, Wu D (2015) Measuring gender-transformative change: a review of literature and promising practices. CARE USA for World FISH and the CGIAR Research Program on Aquatic Agricultural Systems

Hickel J (2015) The problem with saving the world. Jacobian, New York. https://www.jacobinmag.com/2015/08/global-poverty-climate-change-sdgs/. Accessed 15 Nov 2017

Htun M, Weldon L (2012) The civic origins of progressive policy change combating violence against women in global perspective, 1975–2005. Am Polit Sci Rev 106(3):548–569. http://journals.cambridge.org/abstract_S0003055412000226. Accessed 27 Nov 2017

Kaltenborn M, Kuhn J (2017) SDGs acceptance not enforcement. D+C e paper

Karlsson-Vinkhuyzen S, Vihma A (2009) Comparing the legitimacy and effectiveness of global hard and soft law: an analytical framework. Regul Gov 3(4):400–420

Koehler G (2016) Tapping the sustainable development goals for progressive gender equity and equality policy. Gender Dev 24(1):53–68

Marone H (2016) Demographic dividend gender equality and economic growth: the case of Cabo Verde. IMF Working Paper WP/16/169

Mazur A, Hoard S, McBride D (2012) Women's movement change: conceptualization, measurement and investigation. APSA 2012 annual meeting paper. Available at SSRN. https://ssrn.com/abstract=2107885

McArthur J (2013) Own the goals: what the millennium development goals have accomplished. Foreign Aff 92:152–162. https://www.foreignaffairs.com/articles/2013-03-01/own-goals. Accessed 14 Nov 2017

Merry S (2016) Cultural dimensions of power/knowledge: the challenges of violence against women. Sociologie du Travail 58(4):370–380

Nayana B, Das N (2017) Women's inheritance rights, household allocation and gender bias. Am Econ Rev 107(5):150–153

Nayyar D (2012) The MDGs after 2015: some reflections on the possibilities. Paper prepared for the UN System Task Team on the Post 2015 UN Development Agenda

O'Dell K, Peters S, Wharton K (2014) Women, energy and economic empowerment. Deloitte University Press

Onaran O (2015) The role of gender equality in an equality led sustainable development strategy. Greenwich Political Economy Research Center. No: GDERC26
Patchell J, Hayter R (2013) Environmental and evolutionary economic geography: time for EEG2? Geografiska Annale: series B. Hum Geogr 95(2):111–130
Pogge T, Sengupta M (2015) The sustainable development goals (SDGs) as drafted: nice idea, poor execution. Wash Int Law J 24(3):571
Ponte N, Enriquez A (2016) Agenda 2030: a bold enough framework towards sustainable, gender-just development? Gender Dev 24(1):83–98. https://doi.org/10.1080/13552074.2016.1142227. Accessed 10 Nov 2017
Razavi S (1997) Fitting gender into development institutions. World Dev 25(7):1111–1126
Razavi S (2016) The 2030 agenda: challenges of implementation to attain gender equality and women's rights. Gend Dev 24(1):25–41
Romero M (2015) What lies beneath? A critical assessment of PPPs and their impact on sustainable development. Eurodad, Brussels. http://www.eurodad.org/files/pdf/1546450-what-liesbeneath-a-critical-assessment-of-ppps-and-their-impact-on-sustainable-development-1450105297.pdf. Accessed 22 Nov 2017
Rottach E, Schuler R, Hardee K (2009) Gender perspectives improve reproductive health outcomes: new evidence. Population Reference Bureau, Washington, DC
Sharma S, Beaton C, Merill L, Kitson L, Phillip G (2016) Gender and fossil fuel subsidy reform: implications for India, Bangladesh and Nigeria. In: Green Growth Knowledge Platform (GGKP) Fourth Annual Conference on Transforming Development Through Inclusive Green Growth 6-7 September 2016 Jeju International Convention Center, Republic of Korea
Shelton D (1997) Compliance with international human rights soft law. Stud Transnatl Legal Policy 29:119
Sinha A, McRoy R, Berkman B, Sutherland M (2017) Drivers of change: examining the effects of gender equality on child nutrition. Child Youth Serv Rev 76:203–212
UN (2015) Res 69/313, Addis Ababa Action Agenda for the third international development conference on financing and development. UN Doc A/RES/69/313.2015c
UN Women (2013) A transformative stand alone goal on achieving gender equality, women's rights and women's empowerment: imperatives and key components. UN Women, New York. http://wps.unwomen.org/~/media/files/un%20women/wps/highlights/unw-global-study-1325-2015.pdf. Accessed 15 Nov 2017
UNDP (2017) Gender and Climate Finance Policy Brief 5-WEB
Van Beers C (2000) Labor standards and trade flows of OECD countries. World Econ 21(1):57–73
Wolf J (2011) Climate change adaptation as a social process. In: Ford JD, Berrang Ford L (eds) Climate change adaptation in developed nations: from theory to practice. Springer, Dordrecht. https://link.springer.com/chapter/10.1007/978-94-007-0567-8_2. Accessed 23 Dec 2017

Josephine A. Odera is the Director of ACTIL (Africa Centre for Transformative & Inclusive Leadership). Before her appointment at ACTIL, she was the Regional Director, West and Central Africa, UN Women and then briefly served as Regional Advisor, Leadership & Governance at the United Nations Women Regional Office for East and Southern Africa. Mrs. Odera holds a Doctor of Philosophy (PhD) degree in Conflict Management from the University of Ghent, Belgium having previously obtained a master's degree in Economics, a Bachelor of philosophy degree in Economics and a Bachelor of Arts degree.

She has served in the Ministry of Foreign Affairs and International Cooperation of the Republic of Kenya and also taught at the University of Nairobi, Kenya, worked in Civil Society as a Director of Programmes and most recently as Regional Director in the United Nations. As a government official, Dr. Odera represented the Government of Kenya in various meetings including the UN General Assembly, ECOSOC, African Union, regional trade and cooperation meetings such as

COMESA. Dr. Odera was also a pioneer for the review of Foreign Service regulations as they affected women. She has also served as joint secretary to various bilateral commissions of the Republic of Kenya. She has served as a Senior Program consultant for UNDP in Addis Ababa and has undertaken numerous other consultancies in the Horn of Africa and the Great Lakes region.

Judy Mulusa is a Professor of Economics at Prince George's Community College (USA) and an adjunct professor at Bowie State University (USA). She also serves as a Research Associate for Claret Consulting a consulting firm involved in public private partnerships and infrastructure projects. Although her research interests are broad she has a particular interest in economic growth and development, economics or urban redevelopment and public finance. She has most recently published work on gentrification and the housing market in the United States. Previously Judy held faculty positions in her native country, Kenya at Moi University and at Masinde Muliro University of Science and Technology. She also served as an Economist with the Ministry of Finance. Judy received a B. A. in economics and political science minor from the University of Nairobi in Kenya, an M. A. in Economics from the University of Botswana and a Ph. D. in Economics from Howard University (USA).

Superfluous Workers: Why SDG 8 Will Remain Elusive

Christoph Scherrer

Contents

Abstract In 2015, the United Nations agreed to pursue the Sustainable Development Goal 8 "To promote inclusive and sustainable economic growth, employment and decent work for all." The chapter argues that this goal will not be achieved. The abundance of persons offering their labor power in relationship to the limited demand for their labor stems from the insufficient absorption of peasants set free from their land. In many late industrialising countries most of those who are leaving agriculture do not find gainful employment even at the current junction. In fact, many of the late industrialisers are prematurely de-industrialising. Explanations for the lack of absorption capacity of industries and productive services range from overregulated labour markets to globalisation. On the basis of a comparison between the conditions prevalent among the early industrialisers and present-day late comers to industry and advanced services, the chapter highlights other factors: demographic pressures, restrictions on migration, productivity differentials vis-à-vis the Global North and the few successful late industrialisers, and the constraints on the promotion of industry stemming from neoliberal globalisation. It also points to challenges stemming from the colonial heritage such a lack of societal trust.

C. Scherrer (✉)
Universität Kassel, Kassel, Germany
e-mail: scherrer@uni-kassel.de

M. Kaltenborn et al. (eds.), *Sustainable Development Goals and Human Rights*, Interdisciplinary Studies in Human Rights 5,
https://doi.org/10.1007/978-3-030-30469-0_7

Keywords Decent work · Late industrialisation · Vulnerable employment · Absorption capacity of industries · Neoliberal globalisation

In 2015, the United Nations agreed to pursue many Sustainable Development Goals. Goal number eight reads like this: "To promote inclusive and sustainable economic growth, employment and decent work for all." The Decent Work Agenda is the International Labour Organization's strategic response to globalization (ILO 1999). It highlights key dimensions of fair globalization in the world of work by grouping nearly 200 international labor conventions under the following four headings: (1) full employment (including enterprise creation); (2) respect for basic workers' rights; (3) social protection; (4) social dialogue. The Decent Work Agenda received a strong boost in 2005 when the UN World Summit proclaimed decent work for all to be part of the Millennium Development Goals (MDGs). For the first time, governments formally recognized the achievement of full and productive employment as a key instrument for breaking the cycle of poverty.

While this mainstreaming of labor concerns into the United Nations' agenda can be attributed to the ILO's strategic focus on decent work, the world has not moved closer to the fulfilment of the Decent Work Agenda. In fact, the decent work deficit has actually grown. Not only has unemployment increased, but income inequality and informal employment have also been on the rise (ILO 2017).

Since the decent work deficit is more pronounced in the Global South (see Fig. 1), my contribution will focus on the factors responsible for this prevalent deficit among late industrializing countries. I will argue that it is mainly the result of a structural oversupply of labor. The abundance of persons offering their labor power in relationship to the limited demand for their labor stems from the insufficient

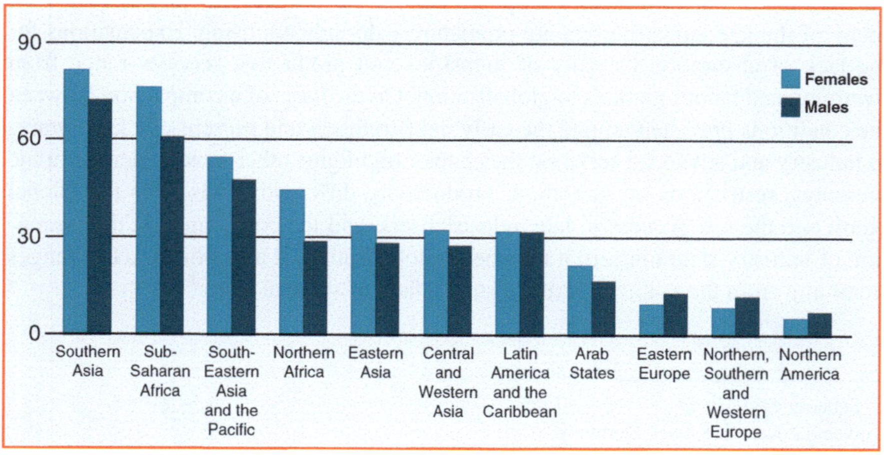

Fig. 1 Vulnerable Employment rates, by sex and regions, 2016 (percentages). Source: World Employment and Social Outlook: Trends 2017; International Labour Office—Geneva: ILO, p. 26

absorption of peasants set free from their land. It leads to what Herbert Gans once called the "superfluous workers" (2012).

Given the low income elasticity of demand for agricultural products, increases in material wealth require the movement of labour out of agriculture. This process has reached a point in the United States where presently, only about 2.1% of the male and 0.8% of the female working population is engaged in agriculture. This compares to about 40.1% male and 60.6% female employment in India and 46.6% and 39.2% in Ghana in 2016 (World Bank 2018). One can imagine how many people would have to leave agriculture in these two countries, if they would reach the US level of productivity. Fortunately, this will take some time not the least because the invested capital per person in agriculture would have to be increased by a factor of 165 (Chen 2016, p. 9). However, in many late industrialising countries most of those who are leaving agriculture do not find gainful employment even at the current junction. In fact, many of the late industrialisers are prematurely de-industrialising. So most of the rural migrants end up in low productivity, low value-added personal services sectors such as petty trade in the informal economy (Dasgupta and Singh 2006; Breman 2013, p. 5).

Explanations for the lack of absorption capacity of industries and productive services range from overregulated labour markets (de Soto 1989, for a critique see Breman 2003, pp. 194–220) to globalisation (Rodrik 2016; his argument will be elaborated below). While the latter explanation has some salience, I want to take up the challenge of the doyen of the study of labour market informality, Jan Breman:

> the research promoted on the informal sector of developing countries from the early 1970s onwards is hampered by the virtual lack of comparison with the profound restructuring from an agrarian-rural to an industrial-urban workforce that went on in the western part of the world at an earlier stage. (Breman 2013, p. 27)

I will carry out a similar comparison between the conditions prevalent among the early industrialisers and present-day late comers to industry and advanced services. My argument takes off from the work of Gavin Kitching (2001) and adds insights from critical development studies. In particular, I will highlight the constraints on the manufacturing sector, especially in sub-Saharan Africa, stemming from the colonial heritage and current global economic governance.

I will start with outlining the current challenges for the Global South's labour markets to provide for sufficient gainful employment. I will move on to develop a framework for explaining these challenges. Based on this framework, I will first elaborate on the demographic pressures on the labour markets followed by a discussion of the factors that limit the capacity of late industrialising countries to accommodate the demand for employment opportunities outside agriculture: restrictions on migration, productivity differentials vis-à-vis the Global North and the few successful late industrialisers, and the constraints on the promotion of industry stemming from neoliberal globalisation.

1 Widespread Vulnerable Employment

In most 'developing' countries, the labour force moving out of agriculture is not absorbed into formal employment in industrial and service sectors. Instead, they move mostly into the informal service sector (Newman et al. 2016, p. 13). In Africa, only about one in five workers has found employment in industry after leaving agriculture (McMillan and Harttgen 2014, p. 2). Overall only 3.2% of the total sub-Saharan workforce was employed in the formal industry in the early 2010s (Losch 2016, p. 15). Many of those who stay behind in rural areas face severe hardship (FAO 2016, p. 14). This resulted in high rate of vulnerable employment especially in southern Asia and sub-Saharan Africa (see Fig. 1).

This is even true for countries which have seen accelerated economic growth in the first decade of the third millennium. The growth champions in Latin America experienced premature deindustrialisation, the ones in Africa barely maintained their earlier low level of manufacturing activities. That these countries were nevertheless growing fast was explained by a team led by Dani Rodrik. In the Latin American case, growth was driven by a commodity boom and manufacturing employment was hit on the one hand by overvalued exchange rates and by labour productivity increases on the other. A significant amount of workers were forced out from high productivity sectors into low productivity activities. The African growth champions profited from the same commodity boom, remittances, and productivity increases in agriculture. The labour force shifting out of agriculture, however, was not absorbed in a dynamic 'modern' sector; hence, the overall labour productivity in the non-agricultural sector declined (Diao et al. 2017). In Ghana, for example, the boom in mining and oil extraction created only a few more jobs in these sectors (Baah-Boateng 2015).[1]

The recent growth champions' experiences differ from the rapid export-oriented industrialisation of East Asian countries such as South Korea, Taiwan, and China. They are, therefore, considered to be less sustainable. If productivity does not increase in the non-agricultural sectors of the economy, then overall growth will be limited (Diao et al. 2017). As the service sector in Africa has absorbed workers faster than the rate of increase of its output, its relative productivity advantage vis-à-vis the rest of the economy has diminished (Newman et al. 2016, p. 11). Manufacturing seems to be better suited to stimulate productivity increases rather than the service sector for catching up economies. The formal sector manufacturing can absorb large number of relatively unskilled workers (i.e. those coming out of agriculture), allows for learning by doing and provides for spill-over effects into the rest of the economy (Rodrik 2013).

However, even successful catching up countries like China have reached their key share of manufacturing employment at much lower levels of GDP per capita

[1]The shares of manufacturing in GDP of most sub-Saharan countries displaying rapid economic growth in recent times, e.g. Ethiopia, Ghana, Kenya, Tanzania, and Uganda, are well below the predicted values for these countries' levels of income (Newman et al. 2016, p. 9).

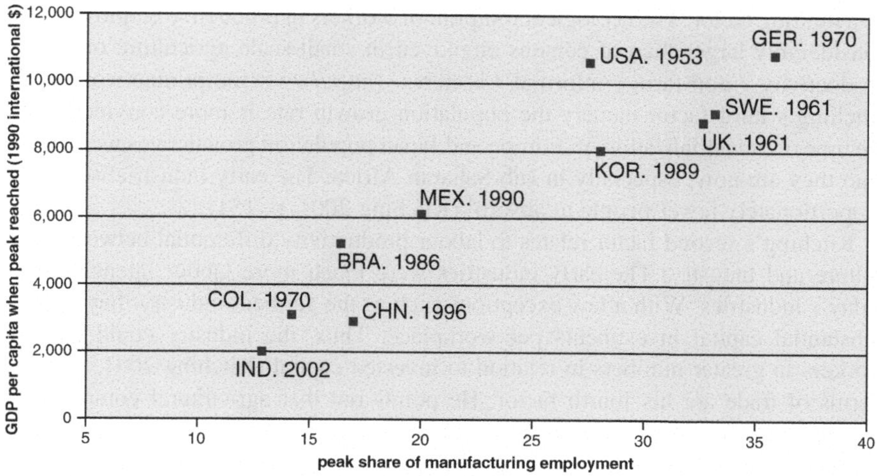

Fig. 2 Peak manufacturing levels, selected countries. Source: Rodrik (2016), p. 20. *BRA* Brazil, *CHN* China, *COL* Columbia, *GER* Germany, *IND* India, *KOR* South Korea, *MEX* Mexico, *SWE* Sweden

than the early industrialisers (see Fig. 2). In addition, the share of low skilled workers employed in manufacturing has decreased across countries of the Global North and South since the late 1990s (Rodrik 2016, p. 19). A recent report on the impact of automation underlines the threat of unemployment in the manufacturing sector in developing countries (Oxford Martin School and Citi 2016). Therefore, the capacity for manufacturing to absorb the rural surplus population seems to be limited. Why is this the case?

2 The Different Contexts of Early and Late Industrialization

To explain the difficulties that countries in sub-Saharan Africa and South Asia experience while moving their agricultural workforce into modern productive sectors, Kitching (2001, pp. 150–52) compares present-day conditions with those when the capitalist core and the Soviet Union had moved from agriculture-based economies to that based on manufacturing. This comparison leads him to highlight five factors that differentiate past experiences from the present ones. His first factor is scale. The rural population of China and India is much larger than what it had been even in the Soviet Union in the 1920s: 'India and China are each faced with a peasant elimination task that is seven to eight times larger than has ever been achieved in human history' (Kitching 2001, p. 150). I find this reference to absolute numbers not so convincing; it has to be qualified in terms of geographical size of the country. From an ecological perspective, however, the absolute size of the population could be a limiting factor. The ecological footprint of workers in productive employment is considerably larger than of persons employed in small-scale agriculture or in low productivity non-farm informal sectors (http://www.footprintnetwork.org/). Kitching's third factor namely the population growth rate is more convincing. At the time of industrialisation, in Europe and Japan population growth rates were lower than they are now, especially in sub-Saharan Africa. The early industrialisers 'had proportionately fewer people to absorb' (Kitching 2001, p. 151).

Kitching's second factor relates to labour productivity differential between agriculture and industry. The early industries were much more labour intensive than today's industries. With a few exceptions such as the garment industry they require substantial capital investments per workplace. Thus, the industry could employ workers in greater numbers in relation to invested capital (Kitching 2001, p. 151). Terms of trade are his fourth factor. He points out that agricultural commodities enjoyed better terms of trade vis-à-vis non-agricultural commodities, i.e., prices for agricultural products went up in relation to prices for industrial goods. These better terms of trade were slowing down the process of 'peasant elimination' because 'those who chose to stay on the land can earn a reasonable living just because prices for the produce are good' (Kitching 2001, p. 152). In contrast, during most of the post-war period, prices for agricultural goods declined in relationship to manufactured goods, therefore, earning a living in small-scale agriculture was difficult. Outmigration becomes more likely and the nonfarm labour market has to absorb proportionately more persons looking for employment (Kitching 2001, p. 152). Kitching's final point highlights different types of crops produced in Europe in comparison to crops in tropical or subtropical regions. However, he does not much elaborate this argument and it seems to me that rice, nuts, fruits, and stimulants produced in the tropics are actually more labour-intensive than growing of grain in temperate climate zones (Khan et al. 2004; Bray 1986).

Kitching sums up his argument: 'neither the contemporary industrial technology context, nor the population growth context, nor the price or terms of trade context, is anywhere near as conducive to peasant elimination as it was when the European world accomplished its (demographically much smaller) transformation' (Kitching 2001, p. 152).

While Kitching focuses more on the labour supply side, Rodrik (2016) analyses the demand conditions for labour, i.e., the limits of employment growth in manufacturing and high value-added service sectors in many of the late industrialising countries, especially in Latin America and Africa. He argues on the basis of extensive analytical statistics that manufacturing employment and output stagnated or even declined once these countries liberalised their trade policies. According to him, 'those without a strong comparative advantage in manufacturing became net importers of manufacturing, reversing a long process of import-substitution' (Rodrik 2016, p. 4). In addition, they were exposed to the decline in relative price of manufacturing caused by technological progress and the rise of Asian exporters. The latter's success came mostly at the expense of other late industrialisers. Particularly hard-hit were the low skilled workers (ibid. 4–19), i.e., those who are most likely from the rural background.

Rodrik also speculates about the political ramifications of premature deindustrialisation. The lack of mass manufacturing comes with a fragmented workforce that is not able to extract from the countries' elites political participation and welfare measures (Rodrik 2016; Breman 2013, p. 7).

My approach builds on the insights of Kitching and Rodrik. It provides further evidence for the arguments concerning population pressure and productivity differential. It also goes beyond the two authors and takes a leave from the pages of critical development studies. In particular, I will highlight the constraints on the manufacturing sector, especially in sub-Saharan Africa, stemming from colonial heritage and current global economic governance.

3 The Democratic Challenge

As I argued earlier, the absolute size of rural population is of less concern as it needs to be seen in relation to the size of the territory. What matters, however, is growth rate. A higher population growth rate requires a faster absorption capacity of manufacturing and higher value added services.

Some development economists have called high growth rates a blessing for respective countries as they would reap a so-called 'demographic dividend'. The dividend would result from a favourable ratio of working age population to children and retired persons, that is, savings from having few dependents would allow for higher capital investments (Lee und Mason 2006). As Adair Turner has pointed out, however, the dividend is dependent on a simultaneous significant fall in fertility. A smaller family size leaves that generation with a larger capital stock per capita and more resources for investment in workforce skills (Turner 2017).

Unfortunately for Africa, its high population growth rates are not accompanied with significantly fewer children per woman. Rural fertility rates controlled for population density are on an average two children higher than other countries of the Global South. This difference is less a result of a desired number of children but more of 'unmet contraception needs' for women (Headey and Jayne 2014, p. 29). In the 1980s and 1990s, China benefited from having two economically active persons for every one inactive person, while sub-Saharan Africa had a ratio of one for one. With the combination of higher fertility rates and an aging population, Bruno Losch is sceptical whether sub-Saharan Africa will even come close to the previous Chinese ratio (Losch 2016, p. 18).

Despite the one-child policy, rapid population increase remains a major labour market challenge for China (Chen and Hamori 2014). How does it compare to the experience of the early industrialisers? Kitching puts the population growth rate for Europe and Japan during their industrialisation phase at roughly 1.5–2% per annum (at the peak), while for the developing countries in the 1990s at 2.5% or 3% and over (Kitching 2001, p. 151). Figure 3 visualises different population dynamics during Germany's industrialisation phase (ca. 1850–1900) and present-day India.

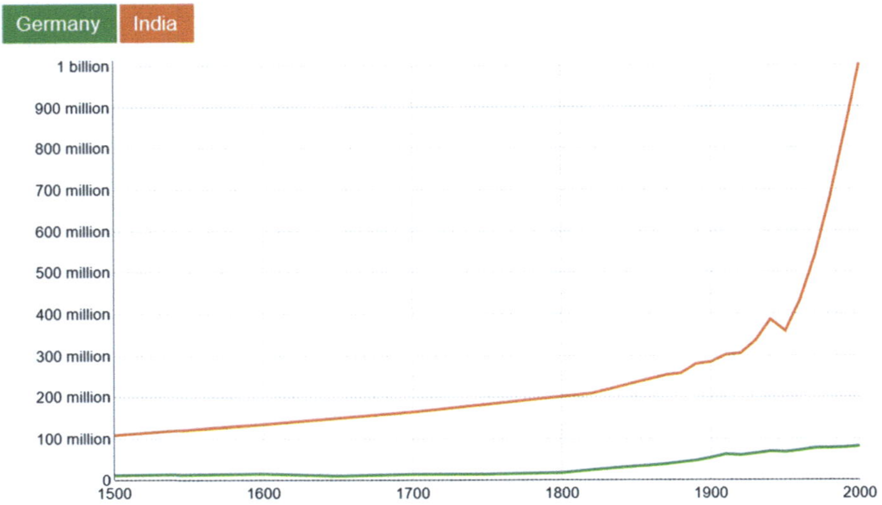

Fig. 3 Population growth Germany and India, 1500–2000. Source: Roser and Ortiz-Ospina (2017)

The birth rate in Germany per 1000 people in the population was on average about 38 in the years between 1850 and 1900; India reached almost a similar rate in 1971 in India, but thereafter moved down to approximately 22 in 2010. As at the time, a high birth rate went along with a higher rate of infancy mortality (Roser 2016); the higher birth-rate in industrialising Germany did not lead to a population growth higher than that in independent India.

Next to a higher birth rate an increase in life expectancy drives population growth. Higher nutritional standards and medical progress have led to a quicker increase in life expectancy in the last decades in comparison to the nineteenth century (Das and Pathak 2012, p. 3). In Germany, life expectancy increased from 41 years to 47 years between 1820 and 1900, and in India from 32 years to 60 years between 1950 and 1999 (Maddison 2001, p. 30; see Fig. 4).

Women's lack of employment in manufacturing has been a cause for high fertility rates. As the experiences in Bangladesh and Lesotho demonstrate, employment of young women in the garment industry makes them more likely to enter school, to stay in school longer, and to postpone marriage and childbirth (Newman et al. 2016, pp. 19–20). The insufficient growth in manufacturing employment aggravates the labour market absorption challenge.

4 Migration: The Narrow Safety Valve

The labour markets of early industrialising countries were relieved from population pressure partly due to massive outflow of people to areas which were less populated in temperate climate zones. After 1815, around 70 million Europeans settled

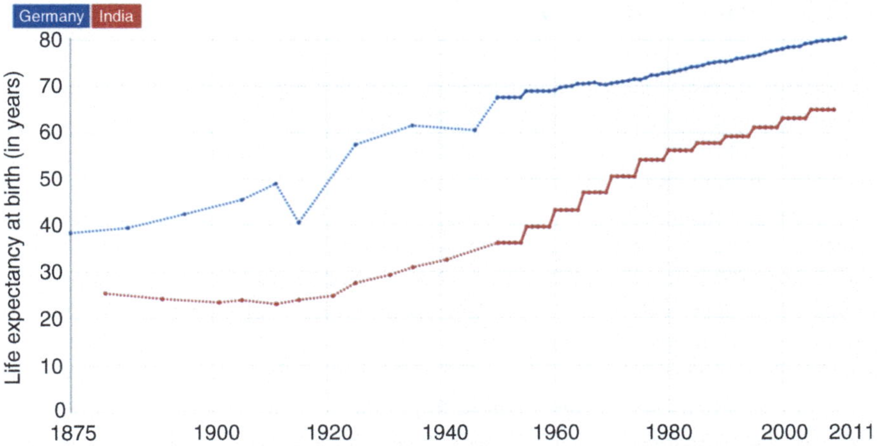

Fig. 4 Comparison of life expectancy at birth, Germany and India, 1875–2011. Source: Roser (2017)

overseas and in Siberia under the umbrella of the military might of the colonial powers or the newly independent white settler republics. On the British Isles and in Norway, mass emigration amounted to more than 30% of their respective populations (Stalker 1994, p. 16). According to Hirst and Thompson, this migration was three times as high as in the 1990s when measured as a portion of the world's population (Hirst et al. 2009, p. 24). Even if these authors might have undercounted the internal migration within large countries such as Brazil, China, and India, the numbers show that for countries which underwent industrialisation later, the outmigration safety valve was and still is much narrower. Most importantly, the migrants have currently to rely on the goodwill of the receiving countries or have to live there on the margins as persons who have violated the migration laws. Unlike the nineteenth century predecessors, they cannot force their way into other territories.

Because of the selectivity of the host countries in contemporary times, emigration is biased towards more qualified persons. Hence, 60% of immigrants from Egypt, Ghana, and South Africa to the United States had a tertiary education in 1990 (Carrington and Detragiache 1998, p. 14). This means for many countries in the Global South there has been a drain of educated people. It is estimated that in recent decades a third of Africa's skilled professionals emigrated (Tanner 2005, p. 3). While this outmigration reduces the pressure on the labour market on the one hand, the loss of so many qualified people, on the other hand, limits the capacity to build a modern economy. It amounts to an educational subsidy for the employers in rich countries.

5 The Productivity Gaps

The labour market for late industrialisers face challenges stemming from three productivity gaps—between the smallholder farmers and modern manufacturing; between smallholders and modern agriculture; and between informal manufacturing and formal manufacturing sectors.

The early industrialisers benefited from more or less simultaneous productivity advances in industry and manufacturing. As industrial technologies were much more labour intensive than today, the industry had a great demand for labour in agriculture. Even in many countries of Asia and Latin America, productivity advances in agriculture were followed by employment increases in manufacturing until the point at which manufacturing's share of total employment reached its peak (Diao et al. 2018, p. 29). However, as the relative importance of manufacturing reached its zenith in these countries at a much earlier date than the early industrialisers (see Fig. 2), the absorption powers of manufacturing were exhausted before the process of 'peasant elimination' had run its course.

In Africa, the productivity gap is even more pronounced (Diao et al. 2018, p. 29). Brazil and China have increased land and labour productivity, but the total factor of productivity for agriculture in sub-Saharan Africa increased by less than 1% per annum (McMillan and Harttgen 2014, p. 14). Among the reasons for the laggard productivity is the diminishing responsiveness to fertiliser use due to over exploitation of land, less use of fertilisers, less conducive conditions for irrigation (in comparison to Asia), greater diversity of crops, underinvestment in crop research (Headey and Jayne 2014, p. 20), and the relative neglect of aid to agriculture over the last three decades by donors (Addison 2017, p. 133).

The large gap between productivity levels of smallholders in Africa and modern manufacturing not only results in a massive labour surplus, but also perpetuates low rural income levels. Low incomes mean low levels of consumption power for industrial products which in turn retards the development of manufacturing. At the time of industrialisation of the North, the smaller gap in productivity advances between agriculture and manufacturing translated into better terms of trade for agricultural products vis-à-vis industrial goods. The relatively higher prices for agricultural goods made the population living off agriculture consumers of industrial products and, thereby, stimulated industrial development. In addition, as agriculture was relatively lucrative and industry developed dynamically, 'peasant elimination' proceeded at a comparatively 'moderate pace' (Kitching 2001, p. 151).

Figure 5 shows substantial land and labour productivity differences among regions of the world. African agriculture, still dominated by more or less self-sufficient smallholders, lags way behind in agricultural output per hectare and worker. While land productivity increased somewhat, labour productivity hardly increased between 1961 and 2009. This gap leaves African agriculture vulnerable to global competition and makes smallholders' land attractive targets for agricultural investors operating on a large scale. In addition, the resulting low incomes make

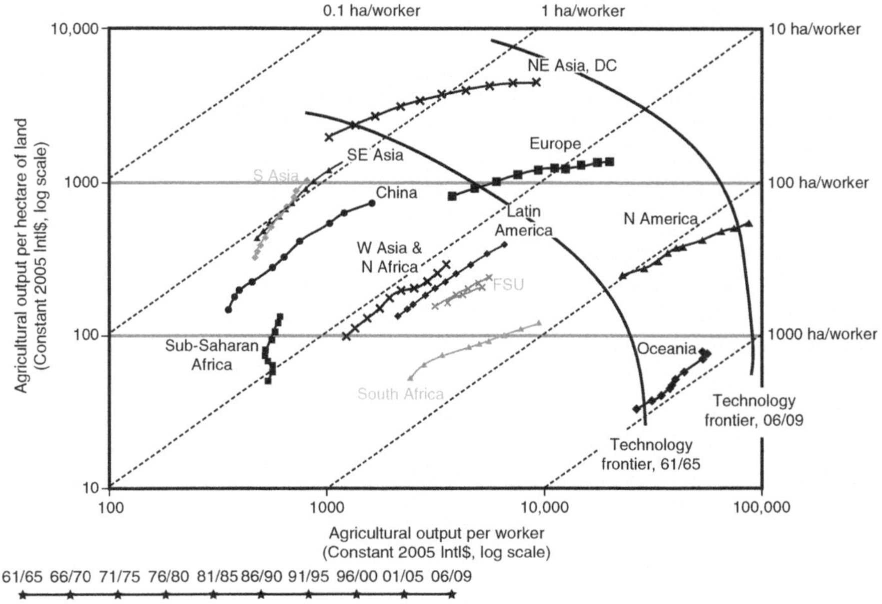

Fig. 5 Agricultural Land and Labor Productivity, World Regions, 1961–2009. Source: Fuglie et al. (2012), p. 2

farm labour unattractive for the rural youth (Losch 2016, p. 46). This productivity gap is, therefore, a major source for the mass movement into cities.

To the extent that the surplus labour is absorbed in manufacturing, it mostly ends up in the informal sector. One of the reasons for this tendency is that while productivity differentials remain high between countries in the agricultural and service sectors, productivity levels converge in formal manufacturing across countries irrespective of 'geographical disadvantages, lousy institutions or bad policies' (Rodrik 2018, p. 17). In other words, agriculture and formal manufacturing are increasing their productivity at different speeds. Higher speed of manufacturing means much less absorption of rural surplus population than at the time of early industrialisation, when productivity in manufacturing was much lower and more in line with agriculture in their specific countries.

The undercapitalised small, informal firms in manufacturing are also lagging much behind in productivity. Even in high-growth years, productivity levels in African manufacturing did not shrink the gap to the US level (Rodrik 2018, pp. 21–23). Higher productivity levels of formal manufacturing implies that investment in manufacturing and output of manufacturing need to grow fast to be able to compensate employment losses in the much less productive informal manufacturing sector. In other words, employment is currently achieved only at the expense of decent work.

6 Globalization's Constraints

The crisis of Fordism in the Global North led to an ever-increasing outsourcing of routine industrial tasks to the Global South since the 1970s. The recipients of outsourcing are unevenly distributed. While over time many, but certainly not all countries, became integrated into global production systems, only a few managed to capture more of the value produced in these so-called global value chains. These successful East Asian economies share a certain characteristic: the capacity of the state and its leading industrial elites to pursue an industrialisation strategy that makes use of foreign financial resources and industrial know-how more or less on their own terms (Azarhoushang et al. 2015).

The great mass of countries was less successful in managing the interface with dominant Northern governments and transnational corporations. Under the dictates of structural adjustment policies they prematurely opened their markets to not only Northern competitors (Addison 2017, pp. 123–130) but, over time, also to their more successful Southern neighbours.

A classic example is Ghana. Its nascent textile industry of the 1970s was reduced to four major textile companies in Ghana employing less than 3000 persons in 2005. It became the victim of imports of second hand clothing from the North and new cheap clothing from Asia (Ackah et al. 2016, p. 63). While the few successful countries moved into the production of more sophisticated products, many of the other countries, especially in Africa, remained stuck in low-sophistication products which even became less sophisticated (Newman et al. 2016, pp. 23–25). According to Adrian Wood, Chinese exporters lowered the ratio of labour-intensive manufacturing to primary output in other countries by 7–10% and the ratio of exports by 10–15% (Wood and Mayer 2009). Only neighbours close to China are integrated in its manufacturing production chains. They benefit in terms of manufacturing employment from the Chinese success in displacing other countries' exports (Jenkins 2016).

While many countries of the Global South opened their borders for northern products, northern countries were slow in reducing the subsidies for their agriculture. It is estimated that US subsidies reduce West Africa's annual revenue from cotton exports by $250 million a year (Fairtrade Foundation 2015).

As tariffs have been reduced in most countries, the level of protection for enterprises from the early industrialised countries has gone up. This is especially true for the increased protection of intellectual property rights. Intellectual property rights, i.e., patents, trademarks, and copyrights, are predominantly held by corporations residing in the early industrialised countries (OECD 2008). Catching up becomes more difficult, if royalties need to be paid for patents.

Besides the protection of intellectual property rights, branding allows corporations from the Global North to dominate global production networks. Without a large customer base in the Global North and the necessary financial resources for advertising, most southern manufacturers have to accept the lower returns for suppliers. The brands use their control over access to the final consumer to force suppliers to lower their prices year by year (Anner 2015).

Besides the liberalisation of cross-border trade, the liberalisation of financial flows limit the policy space necessary for an industrial catch up. The liberalisation of capital accounts left many countries vulnerable to currency crises and capital flight (Herr and Priewe 2005).

7 Limited State Capacity

One of the reasons why many countries lack the characteristics necessary to profit economically from neoliberal globalisation is the shadow of colonialism. While the legacy of colonialism differs among former colonies, they share the fate of having been pushed forcefully into the so-called old division of labour, i.e., being prevented from moving into manufacturing. The enforcement of such a division of labour between the colonisers and the colonised led to deliberate underinvestment in education and skill formation in colonies. It also limited the possibilities for indigenous elites to participate in modern business. Furthermore, the legacy of colonialism meant for most newly independent countries insufficient state capacity and, therefore, weak industrial policies (Breman 2013, p. 117 ff.). Here is not the space to delve deeper into colonialism's ramifications for economic catch up. It has received substantial attention (cf. see World System literature). But one related aspect of great importance for sub-Saharan Africa has only recently been investigated, i.e., the impact of slavery on the homelands of slaves.

A pioneering study by Nunn (2008) through sophisticated econometric calculations suggests that countries with higher losses of people due to slavery in the fifteenth and through the nineteenth century display lower growth rates in their gross domestic product (GDP) in the twentieth century. A preliminary explanation, among other factors, hints at the resulting low trust between villages and within villages. The warfare and raids by competing villages broke up larger societies into smaller ethnically and linguistically differentiated groups. Within these groups, even family members were betraying each other into slavery out of fear of being betrayed (Inikori 2003). A follow-up study which correlated modern trust measures in ethnic homelands with rates of slave extraction found that higher extraction rates predicted mistrust towards family members as well as towards members of other tribes (Nunn and Wantchekon 2011). Slavery extraction left an imprint on today's literacy rates (Obikili 2016). A recent study which analysed slavery's impact on today's access to finance in sub-Saharan Africa provides further support to the claim that in high slave extraction countries, levels of trust are lower than in countries that have suffered less from slavery. The study findings are that firms in such countries not only rely less on formal means of credit but also have less access to informal sources of credit such as from suppliers and customers (Pierce and Snyder 2018).

8 Conclusion: Decent Work Remains Elusive

The extent of vulnerable employment in the Global South is disturbing. But even more disturbing is the prospect that it is likely here to stay if no drastic change happens in the governance of world economy and modes of production as well as consumption. The reason is that the labour market dynamics of the early industrialised countries and the few successful imitators are not easily replicable for all countries. Before I summarise the limiting factors for the large-scale absorption in modern industry of people made superfluous in agriculture, let me state a rather obvious fact which, however, is hardly mentioned in development literature. The industrial development in today's capitalist centres did not only rest on colonial violence but also produced ferocious class struggles and even more devastating wars among the leading industrialisers.

The analysis of the current labour market challenges of late industrialising countries has shown that their industrialisation process takes place under different circumstances. The demographic pressure is significantly more pronounced since fertility rates are not falling quickly enough to compensate for the much quicker increases in life expectancy compared to early industrialisers. The rapid productivity increases in the formal manufacturing sector across the globe limit its absorption powers. Even successful late industrialisers reach the peak of manufacturing's share in total employment much earlier than the first movers of industrialisation. The labour market relief available to these first movers, i.e., outmigration into less densely populated areas, is no longer accessible. Today, migrants cannot overrun indigenous populations with a colonial power backing them up; they have to ask for permission or, if denied, their unlawful presence has at least to be tolerated.

Some countries, especially in Southeast Asia, have partially succeeded in overcoming these constraints. In contrast to many African nations, they had paid more attention to increasing income in agriculture and to overcome infrastructural bottlenecks (Addison 2017). These different strategies have to be seen in the context of diverse colonial and Cold War legacies. The success of some of the South East Asian countries, however, restricts the opportunities for industrialisation for most countries of the Global South. It is a success that rests on massive export surpluses in goods. Yet, the rules governing the world markets limit the value capture also of these successful countries. By strengthening the protection of intellectual property rights and liberalising financial flows across borders, these rules buttress the power of corporations mainly domiciled in the Global North. In competition with each other and faced with high profit expectations from the financial markets, these corporations are dictating the prices of the goods they source from their suppliers.

While rather successful late industrialisers were able to impose some conditions on the business operations of transnational corporations (such as local content requirements and knowledge transfer; Azarhoushang et al. 2015), many other countries lack this capacity due to the shadow of colonialism and, in the case of a number of African countries, due to the detrimental effects of the centuries-long slave extraction on the level of societal trust.

So what are the ramifications of the limited absorption capacity of the modern sector? They boil down to 'superfluous' workers. The oversupply of the working age population severely limits the possibilities for reaching the sustainable development goal number eight. Thus, creative solutions are required on a large scale. Some of the solutions have to be pursued in the Global South, for example agricultural policies that increase rural household income, industrial policies that facilitate diversified economies, and the removal of infrastructural bottlenecks. Other solutions are the responsibility of the Global North, for example restraining the exploitative behaviour of its transnational corporations, changing the rules of global trade and finance in favour of more policy space in the countries of the Global South, and, most important, moving to more sustainable production modes and lifestyles. All countries should strive to distribute work more evenly among the population, thereby making good on the promises of the industrial age: more free time for everyone.

References

Ackah C, Adjasi C, Turkson F (2016) Industrial policy in Ghana: its evolution and impact. In: Newman C, Page J, Rand J, Shimeles A, Söderbom M, Tarp F (eds) Manufacturing transformation comparative studies of industrial development in Africa and Emerging Asia. A study prepared by the United Nations University World Institute for Development Economics Research (UNU-WIDER). Oxford University Press, Oxford

Addison T (2017) 30 years in Africa's development: from structural adjustment to structural transformation? In: van Bergeijk PAG, van der Hoeven R (eds) Sustainable development goals and income inequality. Edward Edgar, Cheltenham, pp 121–140

Anner M (2015) Stopping the race to the bottom. Challenges for workers' rights in supply chains in Asia, International Policy Analysis. Friedrich Ebert Stiftung, Bonn, pp 1–10

Azarhoushang B, Bramucci A, Herr H, Ruoff B (2015) Value chains, under-development and union strategy. Int J Labour Res 7:153–175

Baah-Boateng W (2015) Unemployment in Ghana: a cross sectional analysis from demand and supply perspectives. Afr J Econ Manag Stud 6(4):402–415

Bray F (1986) The rice economies: technology and development in Asian Societies. Basil Blackwell, Oxford

Breman J (2003) The laboring poor in India: patterns of exploitation, subordination and exclusion. Oxford University Press, New Delhi

Breman J (2013) At work in the informal economy of India. The perspective from the bottom up. Oxford University Press, New Delhi

Carrington WJ, Detragiache E (1998) How big is the brain drain?, *IMF Working Paper*, International Monetary Fund, WP/98/102

Chen C (2016) Technology adoption, capital deepening, and international productivity differences. University of Toronto, Toronto, 22 November

Chen G, Hamori S (2014) A solution to the migrant labor shortage and rural labor surplus in China. In: Rural Labor Migration, Discrimination, and the New Dual Labor Market in China. Springer, Berlin, pp 23–38. SpringerBriefs in Economics

Das D, Pathak M (2012) The growing rural-urban disparity in India: some issues. Int J Adv Res Technol 1(5):1–7

Dasgupta S, Singh A (2006) Manufacturing, services and premature deindustrialization in developing countries: a Kaldorian Analysis, *United Nations University Research Paper*, Helsinki: UNU-WIDER, 49/2006

de Soto H (1989) The other path. The economic answer to terrorism. HarperCollins, New York
Diao X, McMillan M, Rodrik D (2017) The recent growth boom in developing economies: a structural-change perspective, NBER Working Paper, 23132, 1–62
Diao X, McMillan M, Wangwe S (2018) Agricultural labour productivity and industrialisation: lessons for Africa. J Afr Econ 27(1):28–65
Fairtrade Foundation (2015) Commodity briefing: Cotton, Accessed 8 Jan 2018 at www.fairtrade.org.uk/~/media/FairtradeUK/What%20is%20Fairtrade/Documents/Policy%20and%20Research%20documents/Product%20related%20documents/Cotton%20Commodity%20Briefing%202015.pdf
FAO - Food and Agriculture Organization of the United Nations (2016) Incorporating decent rural employment in the strategic planning for agricultural development, Accessed 8 Jan 2018 at www.fao.org/3/a-i5471e.pdf
Fuglie KO, Wang SL, Ball VE (2012) Introduction to productivity growth in agriculture. In: Fuglie KO, Wang SL, Ball VE (eds) Productivity growth in agriculture: an international perspective. CAB International, Oxfordshire, pp 1–12
Gans HJ (2012) Superfluous workers. The labor market's invisible discards. Challenge 55 (4):94–103
Headey DD, Jayne T (2014) Adaptation to land constraints: is Africa different? Food Policy 48:18–33
Herr H, Priewe J (2005) The macroeconomics of development and poverty reduction. Strategies beyond the Washington Consensus. Nomos, Baden-Baden
Hirst P, Thompson G, Bromley S (2009) Globalization in question, 3rd edn. Polity, Malden
ILO (2017) World Employment and Social Outlook: Trends 2017, Geneva
ILO - International Labor Organization (1999) Decent Work, Report of the Director-General, International Labour Conference, 87th Session, Geneva
Inikori J (2003) The struggle against the trans-Atlantic slave trade. In: Diouf A (ed) Fighting the slave trade: West African strategies. Ohio University Press, Athens, pp 170–198
Jenkins R (2016) International competitiveness in manufacturing and the China effect. In: Weiss J, Tribe M (eds) Routledge handbook of industry and development. Routledge, London, chapter 15
Khan A, Martin PL, Hardiman P (2004) Expanded production of labor-intensive crops increases agricultural employment. Calif Agr 58(1):35–39
Kitching GN (2001) Seeking social justice through globalization: escaping a nationalist perspective. Pennsylvania State University Press, University Park
Lee R, Mason A (2006) What is the demographic dividend? Finance Dev 43(3):16
Losch B (2016) Structural transformation to boost youth labour demand in sub-Saharan Africa: the role of agriculture, rural areas and territorial development, Employment Policy Department, Employment Working Paper, 204. International Labor Organization, Geneva
Maddison A (2001) The world economy: a millennial perspective. Development Centre of the OECD, Paris
McMillan M, Harttgen K (2014) What is driving the 'African Growth Miracle'?, Working Paper Series, 209. African Development Bank, Tunis
Newman C, Page J, Rand J, Shimeles A, Söderbom M, Tarp F (2016) Made in Africa learning to compete in industry. Brookings Institution, Washington DC
Nunn N (2008) The long-term effects of Africa's slave trades. Q J Econ 123(1):139–176
Nunn N, Wantchekon L (2011) The slave trade and the origins of mistrust in Africa. Am Econ Rev 101:3221–3252
Obikili N (2016) The impact of the slave trade on literacy in West Africa: evidence from the Colonial Era. J Afr Econ 25(1):1–27
OECD - Organisation for Economic Co-Operation and Development (2008) Compendium of patent statistics. OECD, Paris
Oxford Martin School & Citi (2016) Technology at Work v2.0, The Future Is Not What It Used to Be, 8 Jan 2018 at www.oxfordmartin.ox.ac.uk/downloads/reports/Citi_GPS_Technology_Work_2.pdf

Pierce L, Snyder JA (2018) The historical slave trade and firm access to finance in Africa. Rev Financ Stud 31(1):142–174
Rodrik D (2013) Unconditional convergence and manufacturing. Q J Econ 128(1):165–204
Rodrik D (2016) Premature deindustrialization. J Econ Growth 21(1):1–33
Rodrik D (2018) An African growth miracle? J Afr Econ 27(1):10–27
Roser M (2016) Fertility, Accessed 8 Jan 2018 at https://ourworldindata.org/fertility/
Roser M (2017) Life expectancy, Accessed 8 Jan 2018 at https://ourworldindata.org/life-expectancy/
Roser M, Ortiz-Ospina E (2017) World Population Growth, Accessed 8 Jan 2018 at https://ourworldindata.org/world-population-growth/
Stalker P (1994) The work of strangers: a survey of International Labor Migration. International Labour Office, Geneva
Tanner, Arno (2005), Brain drain and beyond: returns and remittances of highly skilled migrants, Global migration perspectives, (24), Geneva: Global Commission on International Migration.
Turner A (2017) The dangers of demographic denial, 29 Sep, Accessed 8 Jan 2018 at www.project-syndicate.org/commentary/developing-countries-demographic-denial-by-adair-turner-2017-09
Wood A, Mayer J (2009) Has China de-industrialised other developing countries, 28 July, Accessed 8 Jan 2018 at www.voxeu.org/article/has-china-de-industrialised-other-developing-countries
World Bank (2018) World development indicators: employment by Sector, Accessed 16 Feb 2018 at http://wdi.worldbank.org/table/2.3

Christoph Scherrer is Professor of Globalization and Politics and the Executive Director of the International Center for Development and Decent Work (ICDD) at the University of Kassel, and a member of the Steering Committee of the Global Labour University. He holds PhDs in political science from the FU Berlin (Dr. habil.) and the University of Frankfurt (Dr. phil.) and a M.A. in economics (U of Frankfurt). He directs two English-language Master programs (MA Global Political Economy (GPE) and Labour Policies and Globalization (LPG)) and one PhD program. He has received the Excellency in Teaching Award of the State of Hessia and the Excellence in Development Cooperation Award from the DAAD.

Reducing Inequality Within and Among Countries: Realizing SDG 10—A Developmental Perspective

Heike Kuhn

Contents

Abstract Respect for human rights is highly relevant for each person, everywhere. At the same time, a closer look is necessary on societies as a whole and their respective levels on inequality. Why? Growing inequality has significant impact on societies and has the potential to undermine democracy. For the first time ever, the global community has agreed upon the goal to reduce inequality within and among countries (SDG 10). This chapter aims to provide an overview of the reduction of inequality from a legal-developmental perspective, discussing the social, economic and ecologic dimension of inequality, the reason behind the highly complex SDG 10, its genesis, the long-standing idea of international solidarity, legal consequences, progress reporting on this SDG, Germany's approach to implement SDG 10 and the road ahead.

H. Kuhn (✉)
Division Human Rights, Gender Equality, Inclusion of Persons with Disabilities, Federal Ministry of Economic Cooperation and Development, Bonn, Germany
e-mail: Heike.Kuhn@bmz.bund.de

M. Kaltenborn et al. (eds.), *Sustainable Development Goals and Human Rights*, Interdisciplinary Studies in Human Rights 5,
https://doi.org/10.1007/978-3-030-30469-0_8

1 A Dedicated SDG on Reducing Inequality: The Relevance of SDG 10

When the 17 SDGs of the 2030 Agenda for Sustainable Development were formulated, the reduction of inequality was made a goal in its own right in SDG 10, which urges global community to "reduce inequality *within and among countries*". A highly difficult and complex goal with seven specific targets, it did, however, find agreement from the 193 UN member states in September 2015 when they endorsed the 2030 Agenda. This fact in itself is surprising and more than a milestone in international cooperation. Why? Because for the first time ever, the common task of reducing inequality within and among countries has been explicitly recognized by heads of states and government and even been combined with a clear timeframe (achievement by 2030).

SDG 10, most interestingly, has ten targets and in them UN member states have found consensus on a broad range of topics relating to inequality, such as

- achieving and sustaining income growth of the bottom 40% of the population at a rate higher than the national average (SDG 10.1),
- empowering and promoting the social and political inclusion of all (SDG 10.2),
- ensuring equal opportunity and reducing inequalities of outcome (SDG 10.3),
- adopting policies and progressively achieving greater equality (SDG 10.4),
- improving the regulation and monitoring of global financial markets and institutions and strengthening the implementation of such regulations (SDG 10.5),
- ensuring enhanced representation and voice for developing countries in decision-making in global international economic and financial institutions (SDG 10.6),
- facilitating orderly, safe, regular and responsible migration and mobility of people (SDG 10.7).

Targets 10.a to 10.c focus on the principle of special and differential treatment for developing countries (SDG 10.a), on encouraging ODA and financial flows to States where the need is greatest (SDG 10.b) and on reducing to less than 3% the transaction costs of migrant remittances by 2030 (SDG 10.c).

This article aims to provide an overview of the reduction of inequality from a legal developmental perspective—the reason behind SDG 10, its genesis, the long-standing idea of international solidarity, legal consequences of SDG 10, progress reporting on this SDG, Germany's approach to implementing SDG 10 and, as a conclusion, the road ahead.

2 Scientific Research on Inequality: The Reason Behind SDG 10

There has been abundant research on inequality in recent decades—showing the harmful effects on human beings when inequality is (too) high within a country. Inequality is seen as one as our most urgent social problems. A worldwide public debate about the 1% and the 99% arose in 2017 when Oxfam drew the following conclusion from the aid figures published by the OECD: just eight men (1%) owned the same wealth as the poorest half of the world.[1] The high level of attention this attracted ignited an intensive debate during the World Economic Forum in Davos in 2017, with leaders from all countries focusing on this topic in numerous debates. A number of recent publications by distinguished economists tackle the topic at length.[2] And the public debate is continuing across the globe, Oxfam remains a strong opinion leader on this topic.[3]

Inequality is—as has been recognized for poverty[4]—multidimensional in nature, including a social, an economic and an ecological dimension.

The Social Dimension of Inequality

As early as 2009, Wilkinson and Pickett[5] provided detailed evidence of the social dimension of inequality, revealing that substantial levels of inequality in income or revenue are indicators of basic problems within societies. The authors show the 'costs' of inequality, with costs being understood in a holistic way and including, for example, mental health and drug use,[6] physical health and life expectancy,[7] obesity,[8] educational performance,[9] teenage births,[10] and higher rates of crime, violence and anti-social behaviour.[11]

This compelling data is drawn from studies undertaken in 25 developed countries. The source was the World Development Indicators Database of the World Bank 2004, which looked initially at the richest 50 countries and subsequently excluded countries with no internationally comparable data on income inequality and those with populations with fewer than three million (in order to exclude tax havens).[12]

Wilkinson/Pickett claim that the inclusion of poorer countries would have made little difference to their results, as studies of life expectancy, infant mortality and

---

[1]Oxfam (2017).

[2]Stiglitz (2012); Piketty (2015); Atkinson (2015); Deaton (2013); Milanovic (2016), Pogge (2015), p. 36 et seq.

[3]Oxfam (2019).

[4]See SDG 1, referring to social protections systems, economic resources and climate-related extreme events; see also Alkire and Foster (2011).

[5]Wilkinson and Pickett (2009).

[6]Ibid., p. 63 et seq.

[7]Ibid., p. 73 et. seq.

[8]Ibid., p. 89 et. seq.

[9]Ibid., p. 103 et. seq.

[10]Ibid., p. 119 et. seq.

[11]Ibid., p. 129 et. seq.

[12]Ibid., p. 280.

homicide in poorer countries show that greater equality is beneficial at all levels of economic development.[13] As this assumption is most persuasive due to the fact that development of societies can be compared on a worldwide scale, it shall be accepted for the purpose of this article.

The Economic Dimension of Inequality

Since humankind has existed, societies have been formed of individuals belonging to higher or lower classes, as reflected in their respective income or assets (inherited or earned by individual work). For more than a century, economists in what is referred to as the Western World declared that economic inequality is inevitable. The main reason given is the underlying assumption that a certain level of inequality is actually most desirable as it provides incentives for entrepreneurs, who invest their capital and knowhow in business, thus creating jobs and wealth for all members of societies.

Today, a more nuanced approach is gaining ground. The World Bank paved the way with its report on 'Poverty and Shared Prosperity 2016: Taking on Inequality', showcasing evidence that there are indeed policies that can help to lower inequality while at the same time boosting growth.[14] The 2018 World Inequality Report, coordinated by several famous economists such as Alvaredo, Piketty and Zucman, also sets a new and different tone, stating that if rising inequality is not properly monitored and addressed, it could lead to various types of political, economic and social catastrophe.[15] These economic experts clearly state that, as it will remain impossible to bring everyone into agreement regarding inequality and no single scientific truth exists about the level of inequality, societies have to decide themselves on the right mix of policies and institutions to achieve the level desired.[16] That decision will be different for a Scandinavian state, say, than for an East African state.

For a long time, societies have struggled to find their own ideal equilibrium. And for all countries worldwide, measuring inequality remains a huge task. Scientists have tried to solve this question with various indices (the most famous ones being *Gini*, *Theil*, *Atkinson* and *Hoover*). At the same time, data are not publicly released in many countries, and surveys tend to underestimate the income and wealth of the richest individuals. Tax havens on all continents and in the middle of oceans mean that some information is not made public.

The World Inequality Report 2018 has delivered some new findings: income inequality varies greatly across all world regions. It is lowest in Europe and highest in the Middle East. In recent decades, income inequality has increased in nearly all

[13]Ibid., p. 281.

[14]World Bank (2016), p. 152 et seq.

[15]World Inequality Report (2018), Executive Summary, p. 4.

[16]Ibid., p. 4.

countries, but at different speeds: since 1980, in North America, China, India and Russia, income inequality has increased rapidly, whereas in Europe it has increased only moderately.[17]

What is most interesting about the World Inequality Report is the intention behind it. The authors aim to measure income and wealth inequality in a systematic and transparent manner, seeking 'to fill a democratic gap and to equip various actors of society with the necessary facts to engage in informed public debates on inequality'.[18] At the same time, it is no surprise that this increased interest in rising inequality has emerged along with globalization and the transparency of the fourth industrial revolution. One could question if this discussion would have been possible without the global agreement reached on SDG 10, evidencing the common ground on which all 193 states were able to agree when arriving at their consensus within the UN in 2015.

The Ecological Dimension of Inequality

In addition to the social and economic dimensions of inequality, there is a third dimension that is of the utmost importance: the ecological dimension. With the Paris Agreement on Climate Change, concluded in December 2015, state parties reached a major breakthrough on adaptation to climate change. The signatory states, having accepted these ambitious goals and targets, will now have to meet them and this will clearly impose strict constraints on enterprises and individuals, forcing many of us to change our comfortable ways. And yet the great majority of the world's population are already being confronted with the ecological dimension of inequality today to a brutal and life-threatening extent: both poor people in rural areas and also the urban poor living in (mega) cities. Threatened by drought, rising sea levels, hurricanes and landslides, poor people and poor countries are suffering most from the climate change, as its impacts affect them disproportionately due to their geographical location and their high socio-economic vulnerability.[19] Looking back, the global warming being experienced today is recognized to be the cumulative effect of the greenhouse gas emissions of the past 200 years. Most of these emissions occurred in the past in the industrialized centres of Western Europe, the USA and the former Soviet Union; it is only in recent decades that prospering Asian states have been emitting a significant proportion of those gases.[20]

[17]Ibid., p. 5.

[18]Ibid., p. 4.

[19]IPCC Working group II (2014).

[20]Scholz (2020), p. 197.

3 The Genesis of SDG 10

When investigating the genesis of SDG 10, it is interesting to look back at the negotiations surrounding the post-2015 Agenda. The negotiations were a highly complex process, involving UN member states, UN agencies, funds and programs, civil society, the private sector and non-state actors. The outcome of these long and difficult negotiations was globally recognized as a major accomplishment, noted by the media in most countries and even inspiring Pope Francis to come to New York and address the General Assembly.[21]

In January 2014, the Permanent Mission of Italy to the UN organized a roundtable to discuss "The Threat of Growing Inequalities: Building More Just and Equitable Societies to Support Growth and Sustainable Development". The intention was to bring together academic knowledge on the impact of growing inequalities and present it to the diplomatic corps; the keynote speaker was the Nobel Laureate in economics Joseph Stiglitz.[22] In his presentation, Stiglitz focused on the level and impact of inequality within the USA, where inequality had increased. Since 2009, he noted, 95% of all economic gains had gone to only the richest 1% of the population. At the same time, he pointed out, the enormous growth in inequality was not only the result of economic forces but also of politics and policies. His conclusion was to call for the SDGs to include a goal to reduce or eliminate inequality in its extreme forms.[23]

Stiglitz' presentation has been described as an "eye-opener". The G77 countries—like others—became aware of and recognized inequality as a global problem.[24] Stiglitz' expertise was trusted, even if it the message was not really new. The academic community,[25] NGOs[26] and even the Pope[27] had, after all, already highlighted high levels of inequality within communities and the impact of it. But in those preparations for a post-2015 development agenda, the message finally got through to the relevant stakeholders.

[21]Kamau et al. (2018), p. 92.

[22]Ibid.

[23]Ibid., 94.

[24]Ibid.

[25]Stewart (2008), p. 3 et seq.; Piketty (2014).

[26]Oxfam (2013).

[27]Pope Francis on World Youth Day 2013, taking place in the Varginha slum of Rio de Janeiro.

4 The Idea of International Solidarity: Nothing New in International Cooperation and Development Policy

For decades, the idea and promise of international solidarity had been discussed intensively in international fora. It is remarkable, but at the same time not surprising, that after World War II states were eager to adopt legal texts and resolutions which explicitly enshrined this principle. However, the concept of solidarity already formed the underlying common ground for the UN Charter in 1945. Article 1 defines as the purpose of the UN, among other things, 'to develop friendly relations among nations based on respect for the principle of equal rights' and 'to achieve international cooperation in solving international problems of an economic, social, cultural, or human character'.

For Europe, the same concept of solidarity is found in the Treaty of Paris of 1951, establishing the European Coal and Steel Community, whose third recital in the preamble states '*recognising that Europe can be built only through practical achievements which will first of all create real solidarity*'. The concept of solidarity was repeated in the Treaty of Rome in 1957, stating in the seventh recital of the preamble '*intending to confirm the solidarity which binds Europe and the overseas countries and desiring to ensure the development of their prosperity'*. And this path was continued in 1992, when the Treaty on European Union was concluded in Maastricht between 12 EU member states, whose heads of state declared their desire 'to deepen the solidarity between their peoples while respecting their history, their culture and their traditions' (fourth recital of the preamble).

It is of interest that Europe, at an early stage, did not only strive for solidarity within its own boundaries, but also with regard to its overseas countries and territories (OCT). Robert Schuman declared in his speech on 9 May 1950, that Europe would, with increased resources, be able to pursue one of its essential tasks: the development of the African continent. Therefore, elements of the principle of solidarity can already be found in the Conventions of Yaoundé[28] and Lomé.[29] This idea of solidarity remained an underlying basis of the subsequent Lomé II to Lomé IV Conventions[30] as well as the Cotonou Agreement of 2000.[31] In preparing the

[28]Yaoundé Convention of 1963 between six European states and 18 African states, 5th recital of the preamble ('résolus à poursuivre en commun leurs efforts en vue du progrès économique, social et culturel de leurs pays').

[29]Lomé I Convention, 1975 between 9 European and 46 African, Caribbean and Pacific (ACP) states, 2nd recital of the preamble ('anxious to establish, on the basis of complete equality between partners, close and continuing cooperation, in a spirit of international solidarity').

[30]David (2000), p. 11 et seq.

[31]Cotonou Agreement 2000, see full text in the Supplement of the ACP-EU Courier, September 2000: Several recitals of the preamble, e.g. second recital: 'affirming their commitment to work together towards the achievement of the objectives of poverty eradication, sustainable development and the gradual integration of the ACP countries into the world economy', third recital 'asserting their resolve to make, through their cooperation, a significant contribution to the economic, social and cultural development of the ACP States and to greater well-being of their population, ... In the

Post-Cotonou Agreement, there is no doubt that this idea will be maintained, considering the history of the Agreements.

Interestingly, the enshrinement of the principle of solidarity took place at a time when decolonization was taking place through the declarations of independence of many colonies. The relationships between colonial powers and their former colonies were often marked by a mutual understanding that a strong bond of solidarity existed—going in both directions and with mutual expectations. That is one of the reasons why the idea of solidarity is inherent in many UN declarations of the 1970s and 1980s. However, it is not until the 1990s, in the UN Millennium Declaration, that the term 'solidarity' can be found in a universally agreed text. The UN Millennium Declaration understands solidarity as one of six fundamental values in international relations: 'Global challenges must be managed in a way that distributes the burdens fairly in accordance with basic principles of equity and social justice. Those who suffer or who benefit least will deserve help from those who benefit most'.

Since 2005, the UN Human Rights Council has appointed an Independent Expert on human rights and international solidarity.[32] That person is mandated to develop a draft declaration on the right of peoples and individuals to international solidarity; that draft was presented back in 2015. The draft declaration on the right to international solidarity takes into account the multitude of international and regional treaties and legal texts that express international solidarity and respect for human rights, stressing that 'international solidarity is a fundamental concept of mutually reinforcing relations among individuals, peoples and States, an essential element that underpins global partnerships, a key approach to peace, disarmament and poverty eradication, and an indispensable component of the efforts to realize all human rights, including the right to development, and internationally agreed development goals'.[33]

5 SDG 10, A New Global Promise of Solidarity: Has It Had Any Legal Consequences?

Coming back to the 2030 Agenda, with SDG 10 and the pledges to leave no one behind and to reach the furthest behind first, it is most relevant to understand the legal nature of this relevant UN resolution. Whereas international treaties are legally binding texts, a resolution does not have the same force. However, resolutions can

effort to give the process of globalization a stronger social dimension', also in referring to the development targets and principles agreed in United Nations Conferences and the OECD Development Assistance Committee to reduce by one half the proportion of people living in extreme poverty by the year 2015 (tenth recital).

[32]The incumbent is Mr. Obiora C. Okafor from Nigeria (since 2017).

[33]Fourteenth recital of the draft declaration on the right to international solidarity.

also reflect and reinforce existing international law. Therefore, UN resolutions are often perceived as "soft law" as they largely reflect and reinforce existing international law.[34] There is a reason for this. Given the complex challenges involved, governments face huge difficulties or even real obstacles in concluding new comprehensive agreements, so in many cases it makes sense to resort to "soft" governance tools. And soft law norms have certain advantages, as they require neither parliamentary approval nor the long time periods of application that customary law is based on.[35]

Taking these arguments into account, the legal character of the 2030 Agenda is of course "soft law" due to the format chosen. It is a UN resolution. But it is a special resolution: dense, intensive, showing consensus among signatory states and many stakeholders involved in the negotiations, providing a framework that can be used as a 'compass' by governments, economists, academia, NGOs, religious entities and many more. A global consensus on the relevance of people, peace, prosperity, planet and peace, promoting the underlying idea of sustainability in all aspects had never been achieved before 25 September 2015, showing the international ambition and the goals, simply "a plan of action for people, planet and prosperity".[36] And solidarity can come from many players: the 2030 Agenda asks for multi-stakeholder partnerships, encouraging effective public, public-private and civil society partnerships (SDG 17.17).

6 Progress Reporting on SDG 10

Not surprisingly, official progress reporting by the UN on SDG 10 is a challenging issue due to the complexity of the target. In its recent 2018 Sustainable Development Report,[37] the progress on SDG 10 is described only in the overview chapter, concentrating on SDG 10.1, 10.a, 10 b. and 10.c, i.e. those targets that can be measured more easily. On SDG 10.1 the report states that, between 2010 and 2016, in 60 out of 94 countries supplying data, the incomes of the poorest 40% of the population grew faster than those of the entire population.[38] More economic data provide information on the progress of SDG 10.a (products exported by LDCs to world markets), 10.b (financial flows to developing countries) and 10 c (transaction costs of remittances).[39] The previous reports have had a similar focus but dealt with the targets in their main chapters, in the 2016 report looking at progress on enhanced

[34]Kaltenborn and Kuhn (2017), p. 17.

[35]Ibid.

[36]Resolution adopted by the General Assembly on 25 September 2015 (70/1) "Transforming our world: the 2030 Agenda for Sustainable Development", preamble, first sentence.

[37]United Nations (2018), p. 9, unstats.un.org/sdgs/files/report/2018/TheSustainableDevelopmentGoalsReport2018.pdf.

[38]Ibid.

[39]Ibid.

representation and voice for developing countries in decision-making in global international economic and financial institutions (SDG 10.6).[40] If one consults the data published on the UNSTATS homepage, the same picture is to be found: SDG 10.1, 10.6, and 10.a. to 10.c are in the focus of interest.[41]

UN Women, not surprisingly, puts the focus on gender aspects and underlines in its stock-taking report of progress towards SDG 10 that, at the national level, gender-responsive fiscal and social policies are needed to reduce income inequalities between women and men, which research shows is a key contributor to overall income inequality in society.[42] A recent study shows that inequality within a household, e.g. between women and men, is a strong contributing factor to overall income inequality in society.[43] We all are witnesses to the fact that in all countries women generally earn less than men, have access to fewer assets and consequently accumulate less wealth. Recent data reveals that across countries, women are more likely than men to be living on less than 50% of the median income.[44]

Taking a look at the reports of UN member states to the annual High-level Political Forum (HLPF), an interesting picture can be seen. In the first round of reporting in 2016, the Synthesis Report was not yet structured to reflect the 17 SDGs (as later reports are).[45] This first ever HLPF asked countries delivering their Voluntary National Reports (VNR) to include a chapter on the principle and the theme of the HLPF in 2016: "ensuring that no one is left behind". The question of reducing inequality was—in a wider sense—discussed in this short chapter, which highlights such aspects as poverty reduction strategies, social policies and social protection. Several European member states, and also Samoa, highlighted the human rights-based approach.[46] A different approach has been taken since the Synthesis Report of 2017. Now the 17 SDGs are shown in different chapters and discussed. Not surprisingly, the issues of social protection policies, social security systems, minimum social standards, human rights and combatting discrimination are again discussed, but also migration policy and enhanced representation and voice for developing countries.[47]

Looking deeper into the experience of Germany (which was an early reporter in 2016), the German government has declared that wealth and income justice, equality and opportunities for participation for everyone are key prerequisites for unlocking

[40]United Nations (2016a), p. 30 et seq., unstats.un.org/sdgs/files/report/2018/TheSustainableDevelopmentGoalsReport2016.pdf; United Nations (2017a), p. 38 et seq., unstats.un.org/sdgs/files/report/2018/TheSustainableDevelopmentGoalsReport2017.pdf.

[41]www.unstats.un.org, SDG 10.

[42]UN WOMEN (2018), p. 113.

[43]Ibid.

[44]Ibid, 114.

[45]United Nations (2016b), https://sustainabledevelopment.un.org/content/documents/126002016_VNR_SynthesisReport.pdf.

[46]Ibid,d., p. 59.

[47]United Nations (2017b), https://sustainabledevelopment.un.org/content/documents/17109_Synthesis_Report_VNRs_2017.pdf.

every person's economic, social and educational potential.[48] According to OECD figures, Germany is among the countries in which government redistribution through taxes and social transfers has reduced income inequality to the greatest extent, lowering the poverty risk by 74%.[49] Four national challenges are mentioned: (1) improving educational equality by creating more opportunities for access and participation for all children and young people across all education sectors; (2) the introduction of a statutory minimum wage; (3) the draft legislation for amending the German Law on Temporary Employment in order to prevent the misuse of work and services contracts; (4) a National Action Plan to implement the UN Convention on the Rights of Persons with Disabilities.[50] In light of its global responsibility, Germany cites three challenges: (1) The German Government advocates for pro-active trade policies which lower tariffs and non-tariff barriers to trade, favoring non-discriminatory trade policy instruments which are conducive to development and mainstreaming of high environmental, labour, social and human rights standards in free trade agreements, trade policy transparency and the participation of civil society stakeholders; (2) the involvement of all social groups (participation); (3) the reform of the World Bank's weighted voting system, underlining that Germany advocates fairer distribution, with voting power being shifted from the industrialized countries to emerging and developing countries with the aim of narrowing inequalities.[51]

Among UN staff, it is understood that reducing inequalities calls for system-wide engagement and must involve all UN entities. It is clearly a major task for the Division for Inclusive Social Development of the Department of Economic and Social Affairs (DESA) to work on the reduction of inequalities, bringing together experts and facilitating dialogue.[52] For an overview of which UN entities are already working on the achievement of SDG 10, the Dalberg Report on "System-Wide Outline of Functions and Capacities of the UN Development System"[53] may be consulted.

And more in-depth analysis of efforts to reduce inequality is still to come. In 2019, SDG 10 will be the subject of an in-depth review by the High-level Political Forum.[54] In preparation of this event, the UN Division for Sustainable Development Goals is planning a number of expert meetings in order to track progress. In 2019, in accordance with the regular 4-year schedule for the composition of delegations, it is

---

[48]https://sustainabledevelopment.un.org, see states and Voluntary National Review of Germany (2016), p. 40.

[49]Ibid.

[50]Ibid.

[51]Ibid., 40 et seq.

[52]See i.a. https://un.org/development/desa/dspd/2018-expert-group-meetings-and-panel-discussions/inequality.html.

[53]United Nations (2017c), https://www.un.org/ecosoc/sites/www.un.org.ecosoc/files/files/en/qcpr/sg-report-dalberg_unds-outline-of-functions-and-capacities_june-2017.pdf.

[54]UN GA resolution 70/299, https://undocs.org/A/RES/70/299.

the heads of state that will convene within the HLPF format. This will draw even greater attention to the session and the progress tracked.

Within Europe, the French Agency for Development has announced it will be staging an International Conference on Development on 7 December 2018, to which experts will be invited for further discussion of the subject (13th AFD International Conference on Development, "Inequalities and Social Cohesion").[55] And the European Commission has recently announced that it will be tackling reducing inequality in its 2019 working programme and will make the topic a prominent focus of the European Development Days in June 2019.

7 Germany's Development Policy on SDG 10

Activities started quite early, with Germany lending its support to the idea of reducing inequality during the post-2015 negotiations, as it was clearly understood that the reduction of poverty and the reduction of inequality are inextricably linked. Once SDG 10 had come into being, the BMZ changed its organizational chart in early 2016 and renamed the division in charge of reduction of poverty the division for "reduction of poverty and inequality".

In September 2016, BMZ organized an International Expert Workshop on Inequalities entitled "Bridging the Gap: Approaches and Policies for Reducing Inequalities". About 80 practitioners and representatives of academia, governments and implementing agencies attended. These discussions led to the *"Expert Key Findings on Reducing Inequalities"*, a two-pager summarizing the main points discussed. The viewpoints expressed are not necessarily shared by all participants nor do they reflect their official positions.[56] Three key messages emerged. Firstly, it was recognised that a strong correlation exists between inequality and conflict and that horizontal inequality threatens peace and stability. Secondly, different distributional outcomes in different countries show that policies matter; political will is the driving force for change, poor governance usually reinforces inequality. Thirdly, development cooperation should be aimed at improving international conditions that cause a rise in inequality within and between countries, i.e. in the areas of trade, taxation and migration. Some of the interesting conclusions to emerge from the workshop were that global rules and policies should be assessed in terms of their effects on inequality; education and health are key to reducing overall inequalities; and, lastly, no group should be left behind and no group should be too far ahead.[57]

---

[55] www.afd.fr./en/international-conference-inequality-and-social-cohesion-2018.

[56] Expert Key Findings on Reducing Inequalities (2016).

[57] Ibid.

Through this workshop, Germany encouraged discussions among the participants on this highly complex topic. And reflections, scientific work, exchange of ideas and dialogue have continued intensively. In addition, a Regional Conference has been taken place in Cape Town (November 2018), tackling the issue of inequality in the southern African region.

Conceptual discussions are still ongoing within the Ministry. Preliminary findings can be summarized as follows:

- Without a significant reduction of inequality, SDG 1 (End poverty in all its forms everywhere) will not be reached.
- Inequality is a driving force for persons to leave their home countries.
- Inequality endangers social cohesion, boosts radicalization and conflict and may have negative effects on economic growth and development successes.
- German development policy aims to look at both ends of the inequality challenge—the most vulnerable and the richest persons of each country—by improving equality of opportunity for all persons globally and enhancing the resilience of poor and vulnerable groups.
- Six main approaches to solving these problems appear to be relevant: political participation and good governance; economic influence (participation) and establishing more just globalization; education and digitalization; gender equality; tax justice; social protection.

The Ministry is also working closely together with all interested stakeholders, creating multi-stakeholder partnerships (as requested by SDG 17.16), involving academia, practitioners and governments in order to exchange knowledge and best practices, an iterative process that has just started and is delivering results. Another interesting step within the Ministry has been the creation of a new fund in 2018, entitled "Seeking new ways – the Inequality Challenge". This fund promotes innovative approaches, methods and tools that focus on reducing inequality and on the poorest and most marginalised people and groups in partner countries of German bilateral development cooperation. It is being implemented by GIZ and has a volume of 1 million euros. The first ten projects have been chosen and will start soon.[58]

8 The Road Ahead

To sum up: By adopting SDG 10, states have shown their political will and commitment to reduce inequality within and among countries. Each country can choose which path it wishes to take towards achieving this ambitious goal by 2030. Reducing inequality within a country can be approached in many ways, depending

[58]See inequality-challenge.com.

on the status quo. That first of all needs to be analysed. Societies suffering from excessive inequality will need to understand that the concentration of capital in the hands of very few people threatens the system in itself, as the various pieces of the puzzle no longer fit together and the consensus needed among the citizens of that society is not guaranteed.

All governments need to establish a framework recognising that all individuals in their countries have the right to enjoy a life in dignity. All human beings, from the moment they are born, need social protection, functioning health services, quality education, decent jobs, reliable institutions and access to modern forms of energy in a safe and healthy environment, as described in the 17 SDGs. Governments have to deliver on those promises for all citizens by guaranteeing minimum standards (e.g. at work, for health, infrastructure, education) and by avoiding a tax system that is seen to tax the average person less fairly than richer members of society. Citizens feel and understand differences in status if inequality among people rises far too much—either through a failure to tax the richest, through bad governance or even corruption, or by guaranteeing privileges only to small parts of societies. And we continue to miss so many talents—clever girls and boys who despite coming from poor backgrounds are brilliant at school, not allowed to continue their education due to the lack of funds.

If individuals or entire groups feel they are excluded from a society, this personal stress causes adverse developments that have become all too familiar: radicalization, high migration numbers, health problems, brain drain, low levels of trust and less willingness to help others. Looked at in a different way: governments will have to realize that if capital is in the hands of very few people, the power this gives to them causes distortions, as governments feel in a weak position with respect to the power behind the capital. Strong lobbyists for the richer minority will favour legal frameworks that guarantee them exclusive advantages (e.g. via tax exemptions, avoiding taxes or subsidies). To a certain extent, governments do react in practice by endeavouring to reallocate, but there are limits. If these reallocations are seen to be unfair, societies face national battles on allocation, with each party stressing their own interests, thus generating high costs and weakening societies.[59] So structural justice in many societies is needed, promoted and implemented by the government and understood by the people. This is a conclusion that applies especially to Germany, which is, according to *Fratzscher*,[60] by a host of measures, one of the most unequal rich countries, if compared globally.

Solidarity among nations is nothing new, as shows the UN Charter of 1945 (Article 1) or—as a regional example—the history of European integration (both among EU member states and also with regard to the EU's relations to its OCTs or

[59]Fratzscher (2016), p. 79 et seq.

[60]Ibid., p. 142 et seq.

developing countries under the umbrella of the Yaoundé, Lomé and Cotonou agreements).[61]

Discussion on this highly controversial topic of reducing inequalities will go on, focusing on the role of governments, individuals, civil society and academia, trying to find ways of achieving a more sustainable and just world by combatting corruption and bad governance, asking for fair taxation for all citizens and establishing liveable societies on all continents.[62] And in order to ensure that all human beings can fulfil their potential in dignity and equality, girls and boys, women and men. With a strong political will and enabling policies, inequality can and will be reduced—if governments have the political will. Having accepted and adopted the 2030 Agenda for Sustainable Development with its SDG 10, 193 governments have declared that they have the political will to act accordingly. Development cooperation that puts much more focus on reducing inequality could ignite discussions on transforming these countries. At the same time, international guidance and the political will of the global community is indispensable. Closing international tax havens is one vital prerequisite for tackling inequality and poverty effectively. Tax evasion undermines the power of governments to provide basic social services including education and health.

In her foreword to Oxfam's report of 2019, Gro Harlem Brundtland, former Director-General of the World Health Organization and first female Prime Minister of Norway, states "Fighting inequality remains one of the world's most testing issues. Delivering universal public services is a tried and tested way to tackle it. We must now take action against extreme inequality to achieve a fairer, healthier and happier future for all, not just the few."[63] Time will tell if the appetite for such efforts exists.

References

Alkire S, Foster J (2011) Counting and multidimensional poverty measurement. J Public Econ 95:476–487; see also https://ophi.org.uk/research/multidimensional-poverty/alkire-foster-method

Atkinson AB (2015) Inequality: what can be done. Harvard University Press, Cambridge

David D (2000) 40 years of Europe-ACP relationship, The ACP-EU Courier, Special issue: Cotonou Agreement, Brussels, pp 11–14

Deaton A (2013) The Great Escape: health, wealth and the origins of inequality. Princeton University Press, New Jersey

Expert Key Findings on Reducing Inequalities (2016) Outcome of the International Expert Workshop on Inequalities "Bridging the Gap: Approaches and Policies for Reducing Inequalities",

[61]The Brexit of Great Britain is, of course, a clear signal of leaving the ideal of solidarity within a community, but does not threaten the idea in itself.

[62]See Martens and Obenland (2017), p. 106.

[63]Oxfam (2019), p. 7.

organized by BMZ, Berlin, 5-6 September 2016, www.giz.de/en/downloads/giz2016-en-expert-key-berlin.pdf (2 pages)
Fratzscher M (2016) Verteilungskampf, Warum Deutschland immer ungleicher wird, München, 2016
IPCC working group II (2014)
Kaltenborn M, Kuhn H (2017) Acceptance, non enforcement – the legal implications of the UN-2030-Agenda and the sustainable development goals. In: Development and Cooperation (D+C), April 2017, https://www.dandc.eu/en/article/2030-agenda-largely-reflects-and-rein forces-existing-international-law
Kamau M, Chasek P, O'Connor D (2018) Transforming multilateral diplomacy, the inside story of the sustainable development goals. Routledge, London
Martens J, Obenland W (2017) Die Agenda 2030 – Globale Zukunftsziele für nachhaltige Entwicklung, Bonn/Osnabrück, pp 99–108
Milanovic B (2016) Global inequality: a new approach for the age of globalization. Harvard University Press, Cambridge
Oxfam (2013) The cost of inequality: how wealth and income extremes hurt us all, Media Briefing 27/2013
Oxfam (2017) An economy for the 99%, 16 January 2017, www.oxfam.org
Oxfam (2019) Public good or private wealth?, January 2019, www.oxfam.org
Piketty T (2014) Capital in the twenty-first century. Harvard University Press, Cambridge
Piketty T (2015) The economics of inequality. Harvard University Press, Cambridge
Pogge T (2015) Wir müssen bei der globalen Architektur ansetzen, Nachhaltige Entwicklungsziele – Agenda für eine bessere Welt?, München, pp 35–40
Scholz I (2020) Reflecting on the Right to Development from the perspective of global environmental change and the 2030 Agenda for Sustainable Development, in this volume, pp. 191–206
Stewart F (ed) (2008) Horizontal inequalities and conflict – understanding group violence in multiethnic societies. Palgrave Macmillan, Basingstoke, pp 3–24
Stiglitz JE (2012) The price of inequality: how today's divided society endangers our future, New York, USA
UN WOMEN (2018) Turning promises into action: gender equality in the 2030 agenda for sustainable development
United Nations (2016a) The Sustainable Development Goals Report, https://unstats.un.org/sdgs/files/report/2016/TheSustainableDevelopmentGoalsReport2016.pdf
United Nations (2016b) Voluntary National Reports (VNR), https://sustainabledevelopment.un.org/content/documents/126002016_VNR_SynthesisReport.pdf
United Nations (2017a) The Sustainable Development Goals Report, https://unstats.un.org/sdgs/files/report/2017/TheSustainableDevelopmentGoalsReport2017.pdf
United Nations (2017b) Voluntary National Reports (VNR), https://sustainabledevelopment.un.org/content/documents/17109_Synthesis_Report_VNRs_2017.pdf
United Nations (2017c) Dalberg report on system-wide outline of functions and capacities of the UN Development System, https://www.un.org/ecosoc/sites/www.un.org.ecosoc/files/files/en/qcpr/sg-report-dalberg_unds-outline-of-functions-and-capacities_june-2017.pdf
United Nations (2018) The Sustainable Development Goals Report, https://unstats.unorg/sdgs/files/report/2018/TheSustainableDevelopementGoalsReports2018.pdf
Wilkinson R, Pickett K (2009) The spirit level – why equality is better for everyone, London
World Bank (2016) Report, Poverty and Shared Prosperity 2016, taking on inequality
World Inequality Report (2018) www.wir2018.wid.world

Heike Kuhn is Head of Division "Human Rights, gender equality, inclusion of persons with disabilities" at the Federal Ministry for Economic Cooperation and Development (BMZ), Bonn/ Berlin). She has a background of more than 25 years in European and multilateral development cooperation, human rights, financing and programming. Ms Kuhn has long-standing experience in leading and managing multilateral negotiations at European and UN-level, lately as Executive Board Director to the International Fund of Agricultural Development and Alternate Representative at the Permanent Representation of the Federal Republic of Germany to the International Organizations in Rome/Italy. Ms Kuhn holds a Ph.D. in Administrative Sciences from the University of Speyer (Germany) on "The social dimension of the European Community". She is a fully qualified lawyer, having studied in Marburg and Freiburg/Germany, Moscow/USSR, Los Angeles/USA and Genoa/Italy.

Securitizing Sustainable Development? The Coercive Sting in SDG 16

Liora Lazarus

Contents

Abstract SDG 16 includes the rule of law alongside a range of factors relating to violence and crime. In this respect, SDG 16 is reflective of a broader shift towards the securitization of the rule of law and human rights. Securitization here refers to the development of concepts to enhance the coercive function of the State. The chapter will demonstrate how SDG 16 thus sits in a broader context of the securitization of the rule of law in international practice, and raise concerns with this trend. The chapter concludes by warning about the potential further securitization of development generally.

1 Introduction

This chapter will argue that SDG 16 represents a culmination, or an exemplar, of a consistent paradigmatic shift—one we can describe as the securitization of the rule of law and human rights. It forms part of a broader academic project concerned with the question of how human rights and the rule of law may be used as legitimating

L. Lazarus (✉)
Bonavero Institute of Human Rights, Law Faculty, University of Oxford, Oxford, UK
e-mail: liora.lazarus@st-annes.ox.ac.uk

M. Kaltenborn et al. (eds.), *Sustainable Development Goals and Human Rights*, Interdisciplinary Studies in Human Rights 5,
https://doi.org/10.1007/978-3-030-30469-0_9

frameworks for coercive overreach of States. Securitization in this context thus implies the paradigmatic shift from the rule of law and human rights as concepts embodying the limitation of the coercive state, to ones integral to the maintenance of law order and security. This is a subtle process and one to which we need to pay more attention in the development process. In this pursuit an exploration of the rule of law's inclusion of Sustainable Development Goal 16, which also incorporates a range of factors relating to violence and crime, is instructive.

2 The Evolving Rule of Law Concept: From Millennium Development Goals to the Sustainable Development Goals

The concept of the rule of law travelled in the fifteen years between the Millennium Development Goals (MDGs) and the Sustainable Development Goals (SDGs), from a mechanism of international justice to one deeply embedded in the fight against failing states. In the MDGs, the rule of law featured in two respects. Under the goal of peace, security and disarmament, strengthening the international rule of law (particularly respect for ICJ decisions) was highlighted as a mechanism for avoiding the 'scourge of war', a concept implicitly understood there as international armed conflict.[1] Under the goal of 'human rights, democracy and good governance' the rule of law was seen also as part of the project of strengthening human rights.[2]

The SDGs have a slightly different view of how 'freedom from violence' is to be realised. Instead of minimising international conflict by strengthening the international rule of law, the SDGs have positioned the rule of law as a tool to achieve the 'supremely ambitious and transformational vision' of 'a world free of fear and violence',[3] a 'world ... in which every child grows up free from violence and exploitation'[4] and in which 'all forms of discrimination and violence against women and girls will be eliminated'.[5] The Rule of Law is thus there 'to foster just and inclusive societies which are free from fear and violence'.[6] This goal is enunciated in SDG 16 which aims to 'promote peaceful and inclusive societies for sustainable development, provide access to justice for all and build effective, accountable and inclusive institutions at all levels'.

One of the key detailed goals under this broader ambition is to 'promote the rule of law at the national and international levels and ensure equal access to justice for all' (16.3). The Rule of law goal is accompanied under SDG 16 by a number of indicators related to violence and crime as:

[1] A/RES/55/2, para 9.
[2] A/RES/55/2, para 24.
[3] A/70/1, para 7.
[4] A/70/1, para 8.
[5] A/RES/70/1, para 20.
[6] A/RES/70/1, 2/35.

inter alia, the goals to 'significantly reduce all forms of violence and related deaths everywhere' (16(1)), to 'end abuse, exploitation, trafficking and all forms of violence against torture of children' (16.2), 'significantly reduce illicit financial and arms flows, strengthen the recovery and return of stolen assets and combat all forms of organised crime (16.4), 'substantially reduce corruption and bribery in all their forms' (16.5), and 'strengthen relevant national institutions, including through international cooperation, for building capacity at all levels, in particular in developing countries, to prevent violence and combat terrorism and crime' (16a). SDG 16 thus undertakes to strengthen the rule of law 'at an international and national level' as a mechanism of 'promoting peaceful and inclusive societies'.

It has introduced the rule of law as a means to reduce violence, terrorism, corruption, trafficking, and serious crime in general at the domestic and transnational level, instead of the focus of the MDGs on avoiding 'international armed conflict'.

A short overview of usage of the rule of law in the United Nations General Assembly (UNGA) indicates that that the approach to the rule of law in the SDG's and SDG 16 in particular, as a mechanism of violence mitigation, as a strategy of shoring up fragile states, is right at the centre of a global emerging consensus. A consensus that views security as a precondition to many things: the good life, human rights, the rule of law, and development. The question is whether all of these concepts can remain intact when premised on this condition. The remaining chapter shows how the rule of law itself has been rearticulated as a mechanism of stability and security over the last decades. This provides a context for the shift between the MDGs and SDG 16, and perhaps as a warning to us.

3 The Rule of Law in International Practice 1993–2016

United Nations General Assembly (UNGA)

The Rule of Law really came prominently onto the international stage at the United Nations World Summit on Human Rights in Vienna in 1993,[7] when the rule of law was highlighted as a key aspect of human rights protection. Ever since this time, the rule of law has operated as a 'strategic and policy envelope' for the United Nations.[8]

In the beginning the emphasis of the Vienna Declaration was mostly on the 'rule of law' as an 'essential factor in the protection of human rights'.[9] Nevertheless, the UNGA 'expressed its conviction that such a programme . . .provide . . . technical and financial assistance to national projects in reforming penal and correctional establishments, education and training of lawyers, judges and security forces in human rights, and any other sphere of activity relevant to the good functioning of the rule of

[7]United Nations, World Conference on Human Rights, 25 June 1993.

[8]Interview with UN official 1, 14 May 2014.

[9]Strengthening the rule of law, A/RES.48/132, 18 February 1994.

law'.[10] Already at this early stage, the institutional scope of rule of law activities included security and criminal justice components.

A line of annual resolutions entitled *Strengthening of the Rule of Law* followed. These sought to make concrete the lofty ambitions of the Vienna Declaration, though continued to highlight the dearth of funds available to do so.[11] Key also was the ongoing recognition that 'rule of law assistance' requests from member States were on the rise.[12]

Alongside, these attempts to embed the mission of the Vienna Declaration, a complementary approach to the Rule of law was developing within the UN. First, the United Nations Millennium Declaration incorporated the rule of law as we have already discussed.

Secondly, and more importantly, in August 2000, the Panel on United Nations Peace Operations reported to the UNGA and the United Nations Security Council (UNSC) on its 'comprehensive review of the whole question of peacekeeping operations in all their aspects'.[13] This report, often referred to under the name of the panel Chairman, Lakhdar Brahimi, contained a range of rule of law proposals. Importantly, one of the key recommendations on peace building was a recommendation for a 'doctrinal shift in the use of civilian police and related rule of law elements in peace operations that emphasises a team approach to upholding the rule of law and respect for human rights and helping communities coming out of conflict to achieve national reconciliation'.[14]

Instead of providing a conceptual definition of the rule of law, the Brahimi report elaborated the notion of 'strengthening the rule of law' by example, in particular 'through training and restructuring of local police, and judicial and penal reform'.[15] When taken together, Brahimi's bricolage of rule of law examples, began to give shape to a more security-centric conception of the rule of law, and significant institutional reform to achieve it.

The Brahimi approach has been hugely influential since its publication. This became clear in 2004 in the SG's report to the Security Council, *The rule of law and transitional justice in conflict and post-conflict societies*.[16] This report enunciated a clear concept of the rule of law for the first time:

> The 'rule of law' is a concept at the very heart of the Organization's mission. It refers to a principle of governance in which all persons, institutions and entities, public and private, including the State itself, are accountable to laws that are publicly promulgated, equally

---

[10]A/RES.48/132, para 2.

[11]A/RES/49/194; A/RES/50/179; A/RES/51/96; A/RES/52/125; A/RES/53/142; A/RES/55/99; A/RES/55/221.

[12]A/RES/57/221 para 3.

[13]A/55/305-S/2000/809.

[14]A/55/305-S/2000/809, page 8, para 47 (b).

[15]A/55/305-S/2000/809, para 13.

[16]UNSC Report of the UN Secretary-General (23 August 2005). The rule of law and transitional justice in conflict and post conflict societies, S/2004/616*.

> enforced and independently adjudicated, and which are consistent with international human rights norms and standards. It requires, as well, measures to ensure adherence to the principles of supremacy of law, equality before the law, accountability to the law, fairness in the application of the law, separation of powers, participation in decision-making, legal certainty, avoidance of arbitrariness and procedural and legal transparency.[17]

This conception has been consistently applied since then. The most recent SG analytical summary of rule of law debates in the General Assembly between 2006 and 2015, described the definition as an 'important milestone in the area', pointing out 'numerous delegations expressed their agreement' with this definition in more recent years.[18]

Certainly, the SG definition includes classical state limiting norms normally espoused within the liberal rule of law canon. Nevertheless, it goes on to include private actors and institutions as subjects of the rule of law. This opens the door to the coercion of private actors and institutions as a necessary part of the rule of law, and engages by definition the criminal justice system as a whole.

It is also important to bookmark here that UN's rule of law definition was first presented within a report, and a forum, which was focused on security concerns. In the context of the UN General Assembly, this security-centric tone echoed sonorously in the 2005 report, *In Larger Freedom*.[19] The report saw larger freedom, development and security, as going 'hand in hand',[20] and emphasised the 'mutually reinforcing' ... 'right to security and to development'.[21] Crucial to these twin aims and the achievement of 'collective security' is the shoring up of fragile states. Given the centrality of State capacity to collective security, the rule of law featured heavily in the report. Firstly, the rule of law featured as part of the 'freedom from want' agenda, by bolstering secure investment in developing economies.[22] Secondly, it featured as a central tool of 'peace-building' and the attainment of 'freedom from fear'.[23] Finally, a whole section was devoted to the rule of law in relation to 'freedom to live in dignity'.

But despite this title, the discussion of the rule of law mostly emphasised security and protection. Indeed, the report declared that the Millennium Declaration's emphasis was on the 'rule of law as the all-important framework for advancing human security and prosperity'.[24] Strikingly, private actors threatening security, such as armed groups or terrorists were cast as those who 'make no pretence of being bound

---

[17]S/2004/616∗, 23 August 2004, para 6.

[18]Strengthening and coordinating United Nations rule of law activities, A/70/206, 27 July 2015, Annex, para 8.

[19]In larger freedom: towards security, development and Human Rights for all, 21 March 2005, A/59/2005.

[20]A/59/2005/ para 15.

[21]A/59/2005, para 16.

[22]A/59/2005, para 36.

[23]A/59/2005, para 115.

[24]A/59/2005, para 133.

by the rule of law'. The phrase sought to describe a juxtaposition between a society following the rule of law and the threat of mass violence.

The *Larger Freedom* report was saturated with images of state collapse, and the violence that flows from a disregard for the rule of law and law in general. Strengthening the rule of law was seen as an answer not to the overreach of State coercion, but rather to the civic violence that erupts inside fragile States, or as a response to the impunity of those individuals responsible for breaches of international criminal law.

Following on from the 2005 World Summit Outcome document, the 'rule of law at the national and international levels' became a permanent annual agenda item on the Sixth Committee of the General Assembly.[25] From this point, SG reports[26] and UN General Assembly resolutions[27] on strengthening the rule of law were an annual feature.

Alongside this annual cycle, the concept of the rule of law has constituted a focus of other themed reports. In 2007, the rule of law as a foundation of security featured heavily in the SG's '*Comprehensive report on strengthening the capacity of the United Nations to manage and sustain peace operations*'.[28] This proposed a series of structural changes to the United Nations Department of Peacekeeping Operations (DPKO). The structural proposal was to divide the DPKO into five components, one of which would be the 'Office of Rule of Law and Security Institutions' (OROLSI).[29]

The 2007 report noted that 'one of the key aspect of the Brahimi report was its stress on the significance of the rule of law and public order for sustainable peace and security in countries emerging from conflict.[30] But it also brought to that message the benefit of 7 more years of peacekeeping experience:

> the lessons of the past decade of peacekeeping have also shown that the establishment and reform of the rule of law in post-conflict contexts is dependent on a basic level of security . . . establishment of national security strategies and institutions operating under the rule of law are essential for the development of judicial, legal, correctional and police reform. This recognition has led to the initiation of comprehensive inter-agency consideration of the role of the United Nations in security sector reform. This exercise, while still ongoing, has underscored the linkage between the rule of law and security institutions.

---

[25]Organisation of the sixty-first session of the General Assembly, adoption of the agenda and allocation of items, First Report of the General Committee, A/61/250, paras 56, 75, 79.

[26]Reports of the Secretary General: the rule of law at the national and international levels, A/62/121, 11 July 2007; Strengthening and coordinating United Nations rule of law activities, A/63/226, 6 August 2008.

[27]General Assembly resolutions on the rule of law at the national and international levels: A/RES/62/70, 8 January 2008; A/RES/63/128, 15 January 2009; A/RES/64/116, 15 January 2010; A/RES/65/32, 10 January 2011; A/RES/66/102, 13 January 2012; A/RES/67/97, 14 January 2013; A/RES/68/116, 18 December 2013; A/RES/69/123, 18 December 2014.

[28]A/61/858, 13 April 2007.

[29]A/61/858, para 26.

[30]A/61/858, para 57.

Hence in 2008, a 'vision of security based on the rule of law' became part of the Secretary General's report on security sector reform.[31] The report reinforced the clearly emerging UN consensus that the rule of law and security were now intertwined, featuring the 'rule of law' as the 'only' framework in which sustainable rights regarding security could be achieved, while also placing security itself as the 'precondition' to human rights.[32]

SG reports and GA resolutions on strengthening the rule of law continued on an annual basis, until the 2012 *Declaration of the High-Level Meeting of the General Assembly on the Rule of Law at the National and International Levels.* This declaration was similarly imbued with a sense of urgency about threats to peace and security. While also emphasising the constraining normative frameworks applying in the development of rule of law capacity, the General Assembly emphasized 'the importance of the rule of law as one of the key elements of conflict prevention, peacekeeping, conflict resolution, peace building' stressing 'justice' and 'transitional justice mechanisms' as key to this end.[33] Similarly, it stressed the 'importance of supporting national civilian capacity development and institution-building in the aftermath of conflict'.[34]

The resolution also reinforced the rule of law's place on the UNDP agenda, by articulating the inescapable connection between rule of law work and a range of development goals, as well as the 'right to development':

> We are convinced that the rule of law and development are strongly interrelated and mutually reinforcing, that the advancement of the rule of law at the national and international levels is essential for sustained and inclusive economic growth, sustainable development, the eradication of poverty and hunger and the full realization of all human rights and fundamental freedoms, including the right to development, all of which in turn reinforce the rule of law, and for this reason we are convinced that this interrelationship should be considered in the post-2015 international development agenda.[35]

Since the High Level Declaration, the UNGA has continued to generate annual resolutions reiterating its commitment to developing the 'rule of law at the national and international levels', and[36] SG reports in response to them.[37] All of these reports stress the relationship between peace, security, development and the rule of law. There are too many to mention, but as an example the *2015 Report of the Special*

---

[31]Securing peace and development: the role of the United Nations in supporting security sector reform, A/62/659-S/2008/39, 23 January 2008, para 4.

[32]On the growing influence of the idea that security constitutes a precondition to human rights see: Lazarus (2015).

[33]A/RES/67/1, para 18.

[34]A/RES/67/1, para 19.

[35]A/RES/67/1, para 7.

[36]A/RES/67/97, 14 January 2013; A/RES/68/116, 18 December 2013; A/RES/69/123, 18 December 2014; A/RES/70/118, 18 December 2015.

[37]A/69/181, 24 July 2014; A/70/206, 27 July 2015.

Committee on Peacekeeping Operations is again indicative of this trend.[38] The report devoted an entire section to the rule of law emphasising on numerous occasions the 'critical importance of strengthening the rule of law in countries in conflict and emerging from conflict in order to help stabilize the situation, extend State authority, end impunity, tackle the underlying causes of conflict and build lasting peace'.[39] Similarly, the report 'underline[d] the important role that peace-keeping operations ... can play in helping national authorities ... to support the initial strengthening of the host State's rule-of-law institutions'.[40] The support of peacekeeping missions for rule of law institutions was a key focus, in particular with respect to early interventions aimed at 'maintaining basic law and order and fighting impunity by strengthening national police, justice and corrections institutions to restore the rule of law'.[41] The range of rule of law actors acknowledged in the report were considerable, and a clear indication of how far rule of law work had become mainstreamed into both organisational reform of the UN itself and the management and structure of peacekeeping missions on the ground.[42]

This survey of UN General Assembly and Secretary General documentation on the rule of law has been purposively selective. Its aim is to highlight the growing emphasis on the conceptual relationship between the rule of law, security, law and order, peacekeeping and development. Undoubtedly, the UN's commitment to human rights, access to justice, and to the normative structures constraining and holding State's accountable to law is clear, consistent in all its publications and embedded in the SG's 2004 definition of the rule of law.

Nevertheless, the reference to 'private' persons, 'institutions and entities' and the idea of 'supremacy of law' provides a conceptual starting point for a broader conception of the Rule of Law as an apparatus aimed at maximising security and law and order, minimising violence, and embedding peace. It is unarguable that there is a continuous stream throughout all these documents linking the rule of law to the extension of State authority and the control of crime in fragile states. There is in short, a coercive sting in the tail of the Rule of Law, and no less so in SDG 16.

---

[38] Report of the Special Committee on Peacekeeping Operations 2015 substantive session, 17 March 2015, A/69/19.

[39] A/69/19, para 170.

[40] A/69/19, para 171.

[41] A/69/19, para 175.

[42] A/69/19, paras 176–185. See also further Sect. 4 below.

United Nations Security Council (UNSC): Rule of Law Vacuums

Let us turn briefly to the Security Council itself, where Rule of Law activity is ubiquitous. Probably the quickest and most instructive way to understand all aspects of the SC's working conception of the rule of law, is to look at its opposite construction: a 'rule of law vacuum' a term commonly invoked in the SC and in peacekeeping:

> In post-conflict settings, legislative frameworks often show the accumulated signs of neglect and political distortion, contain discriminatory elements and rarely reflect the requirements of international human rights and criminal law standards. Emergency laws and executive decrees are often the order of the day. Where adequate laws are on the books, they may be unknown to the general public and official actors may have neither the capacity nor the tools to implement them. National judicial, police and corrections systems have typically been stripped of the human, financial and material resources necessary for their proper functioning. They also often lack legitimacy, having been transformed by conflict and abuse into instruments of repression. Such situations are invariably marked by an abundance of arms, rampant gender and sexually based violence, the exploitation of children, the persecution of minorities and vulnerable groups, organized crime, smuggling, trafficking in human beings and other criminal activities. In such situations, organized criminal groups are often better resourced than local government and better armed than local law enforcement. Restoring the capacity and legitimacy of national institutions is a long-term undertaking. However, urgent action to restore human security, human rights and the rule of law cannot be deferred. Thus, United Nations peace operations are often called upon to help fill this rule of law vacuum.[43]

Inside this 'rule of law vacuum' we see not only a legal system which violates rule of law norms, but also a system of enforcement which is dysfunctional and illegitimate and unable to control the rampant crime, violence and disorder described as 'better resourced' than government actors. The task, then, of rule of law intervention is to restore both the 'capacity' and the 'legitimacy' of 'local law enforcement'. While legitimacy of local law enforcement actors features large in this description, equal attention is given to their capacity to enforce order.

The experience of UNSC oversight of transitional societies after large scale conflict and atrocities over the past decade has reinforced the view that there is a key relationship between peace, stability and the elimination of 'rule of law' vacuums. Consequently, the nexus between security and the rule of law has been consistently emphasised in SC Secretary General reports and letters,[44]

---

[43]UNSC Report of the UN Secretary-General', S/2004/616*, 23 August 2004, para 27.

[44]UNSC Reports of the UN Secretary General: S/2004/616*, 23 August 2004; S/2006/980, 14 December 2006; S/2008/39, 23 January 2008; S/2008/460, 15 July 2008; S/2009/277, 29 May 2009; S/2009/304, 11 June 2009; S/2009/465, 16 September 2009; S/2010/288, 30 June 2010; S/2011/527, 11 August 2011; S/2011/634, 12 October 2011; S/2012/645, 15 August 2012. Letter from the UN Secretary General to the President of the UN Security Council, S/2012/958, 19 December 2012.

SC reports,[45] and SC Presidential Statements,[46] and also evidenced by the establishment of a Rule of Law Assistance Unit within the Security Council.[47]

In the recent 2014 Statement, the Security Council President reaffirmed the 'importance of the rule of law as one of the key elements of conflict prevention, peacekeeping, conflict resolution and peace building'.[48] The Statement went on to note 'the important role that the police components of peacekeeping operations can play in strengthening the rule of law in conflict and post-conflict situations, by, inter alia, providing operational support to national police and other law enforcement agencies and supporting the reform, restructuring and rebuilding of such agencies, including through technical assistance, co-location, training and mentoring programmes'.[49] There is little question therefore, that the language of the first thematic rule of law report has since become ubiquitous in official SC documentation.

In sum, from a closer analysis of their own sources, there is considerable evidence to show that UNSC peace building strategies are interleaved with a securitised conception of the rule of law. What the UNSC terms, 'citizen security' or a 'victim centred approach to peacekeeping', is at the heart of its pursuit of social ordering structures inside unstable societies. The practical operational realities of shoring up such structures has been repeatedly emphasised by the UNSC which has 'faced the difficulties of conducting peace operations where there are no functioning criminal justice mechanisms at all'[50] and that 'lawlessness can seriously undermine the efforts of entire peace operations'. As a consequence, the strategies for filling 'rule of law vacuums' in the first instance particularly include the role, capacities and obligations of military and civilian police components'.[51]

All of these peace keeping strategies, and resources to shore up security and order, are incorporated into the UNSC's framing of the 'the rule of law'. But there is insufficient conceptual attention to the distinction between the rule of law as a tool of security, stability and peace, and the normative ideals of the UNSG's rule of law concept as a state limiting device. The background coercive conditions which are necessary for the preservation of social order, are described as part of rule of law practice and rhetoric over and over again. There isn't a clear, consistent and

[45]S/PV.4833, 24 September 2003; S/PV.5052, 6 October 2004; S/PV.5474, 22 June 2006; S/PV.6347, 29 June 2010; S/PV.6705, 19 January 2012; S/PV.6849, 17 October 2012; S/PV.6913, 30 January 2013.

[46]S/PRST/2003/15, 30 September 2003; S/PRST/2004/2, 26 January 2004; S/PRST/2004/32, 10 September 2004; S/PRST/2005/30, 12 July 2005; S/PRST/2006/28, 22 June 2006; S/PRST/2010/11, 29 June 2010; S/PRST/2012/1, 19 January 2012; S/PRST/2013/1, 15/01/2013; S/PRST/2013/4, 15/04/2013; S/PRST/2014/3, 12 February 2014; S/PRST/2014/4, 14 February 2014; S/PRST/2014/5, 21 February 2014.

[47]Statement by the President of the UNSC, S/PRST/2006/28, 22 June 2006.

[48]Statement by the President of the UNSC, S/PRST/2014/5, 21 February 2014, page 1.

[49]Statement by the President of the UNSC, S/PRST/2014/5, 21 February 2014, page 2.

[50]UNSC Report of the UN Secretary-General', S/2004/616*, 23 August 2004, para 28.

[51]UNSC Report of the UN Secretary-General', S/2004/616*, 23 August 2004, para 10.

concerted effort to distinguish security capacity building as an institutional pre-condition to the realisation of traditional rule of law safeguards. While there is consistent reference to human rights safeguards, justice and other values safeguarded by the rule of law, there is less consistent and explicit acknowledgement of the deep epistemological conflict between the measures aimed at bolstering fragile states and securing citizens, and those aimed at ensuring that the state is limited by rule of law safeguards.

4 The Rule of Law and the United Nations Development Programme (UNDP)

Let us turn now to the role of the UNDP within the Rule of Law architecture of the UN.

The UN now has a huge range of institutions involved in the delivery of rule of law work. At present there are two main bodies which co-ordinate this work at headquarter level. The Rule of Law Coordination and Resource Group (RoLCRG),[52] which is chaired by the Deputy Secretary General, has the overall leadership role for rule of law work. The RoLCRG is made up of nine members, including the UNDP and DPKO, which in turn are responsible for 20 organisations in the delivery of rule of law assistance.[53]

The second important institution steering the Rule of Law in the UN is the ‘Global Focal Point for Police, Justice and Corrections Areas in the Rule of Law in Post-conflict and other Crisis Situations’ (GFP) established in 2012. The GFP brings key institutions together who will oversee crisis conditions on the ground, in particular the DPKO, and the UNDP.[54] The GFP says it is there to ‘support the crisis system in delivering police, justice and corrections assistance to peacekeeping and political mission settings and other crisis situations, and to assist the United Nations country teams and missions to develop and implement common rule of law strategies

[52]https://www.un.org/ruleoflaw/what-is-the-rule-of-law/coordination-of-rule-of-law-activities/ (last accessed 7 August 2019).

[53]The RoLCRG is made up of nine members: the Department of Political Affairs, the Department of Peacekeeping Operations (DPKO), the office of the High Commissioner for Human Rights (OHCHR); the Office of Legal Affairs; the United Nations Development Programme; the United Nations Development Fund for Women, and the United Nations Office on Drugs and Crime. This Group co-ordinates the activities of 20 UN institutions which include the rule of law in their mandates: DESA, DPA, DPKO, OCHA, OHCHR, OLA, OSRSG for Children and Armed Conflict, OSRSG on Sexual Violence in Conflict, PBSO, UNDEF, UNEP, UNESCO, UN-Habitat, UNHCR, UNICEF, UNODC, UNOPS, UN Women, World Bank. The RoLCRG is supported by the Rule of Law Assistance Unit.

[54]The GFP also includes the Department of Peacekeeping Operations, the United Nations Development Programme, the Office of the High Commissioner for Human Rights, the United Nations Office on Drugs and Crime, and UN Women. See further in general: https://www.un.org/ruleoflaw/what-is-the-rule-of-law/coordination-of-rule-of-law-activities/.

and programmes.[55] It is clear from the focus and design of the GFP that rule of law operations in crisis conditions intermesh development, human rights and security activity.

It is clear that the UNDP is part of both major rule of law networks within the UN, both general and in crisis missions. In particular, the UNDP is now deeply integrated into a rule of law partnership with the DPKO inside both of these networks. Consequently, it is not surprising that a peacekeeping aspect to the rule of law has travelled across to the UNDP. The role of the UNDP was outlined by the SG in 2008:

> The UNDP plays an active role in crisis and post-conflict situations and supports the implementation of capacity-development programmes tailored to put in place foundations for recovery and development. UNDP assistance entails both justice and security and is geared towards long-term development and sustainability. At the request of Governments, UNDP emphasises the protection of civilians, access to justice and the rule of law and democratic governance in recovery and peace building environments.

Importantly, we can see from this description, that the UNDP is involved in the delivery of both justice and security—with a particular emphasis on the protection of civilians and on access to justice and the rule of law.

At the outset it must be acknowledged that the UNDP brings a different tone to rule of law interests. It promotes a broader and softer approach to the Rule of Law than the DPKO. It joins up the ‘rule of law’, ‘justice’, ‘security’ and ‘human rights’ in its account of its own work.[56] Thus SDG 16 makes ample reference to ‘fundamental freedoms’ (SDG 16.10), and a range of normative goals (e.g. SDG 16.7—‘ensure responsive, inclusive, participatory and representative decision making’ and SDG 16 b: promote and enforce ‘non-discriminatory’ laws and policies for sustainable development’). Similarly, the UNDP guidance on RoL indicators distinguishes between the ‘often overlapping’ concepts of ‘rule of law’, ‘justice’ and ‘security’.[57]

5 Conclusion: Securitizing Development?

Nevertheless, the intermeshing of peacekeeping and development goals as joint rule of law objectives leads inevitably to an emphasis on objectives which further the urgent need to shore up ‘State authority’ in crisis conditions as well as rights regarding normative structures. This is why SDG 16 includes concrete objectives relating to the reduction of crime and violence:

---

[55]Fact Sheet: Global Focal Point for Police, Justice and Corrections UNDP and DPKO, December 2012. http://www.undp.org/content/dam/undp/library/crisis%20prevention/2012_12_07_The_Fact_Sheet_for_Global_Focal_Point_Dec_2012.pdf (last accessed 7 August 2019).

[56]See UNDP website “Our Work”: http://www.undp.org/content/undp/en/home/ourwork/democratic-governance-and-peacebuilding/rule-of-law%2D%2Djustice-and-security.html.

[57]UNDP, Why, What and How to Measure, p. 3.

> 'significantly reduc(ing) all violence and related death rates everywhere' (16.1), ending "abuse, exploitation, trafficking and all forms of violence against and torture of children" (16.2), significantly reducing "illicit financial and arms flows, strengthen recovery and return of stolen assets and combat all forms of organised crime" (16.4), and "substantially reduc (ing) corruption and bribery in all their forms"'.[58]

These concrete crime and violence reduction objectives indicate that building the rule of law to further development and peacekeeping, goes hand in hand with crime control.

This message is repeated in other UNDP institutional material. The UNDP's rule of law co-ordinating function is led by the 'Rule of Law, Justice and Security Unit', which co-ordinates with OROLSI within the GFP. Its promotional material reiterates the message that the rule of law is the basis for security, stability, peace and development.[59] Its simple message on its promotional video echoes the SC's 'rule of law vacuums' approach: 'when the rule of law breaks down, society crumbles'.[60] The video goes on to feature stabilising objectives such as the reduction of armed violence, the support of SGBV victims, and the promotion of national reconciliation, as successes of joint rule of law efforts between the UNDP and DPKO working on the ground.[61]

Similarly, in the 2014 UNDP *User's Guide to Measuring Rule of Law, Justice and Security Programmes*, the UNDP makes it clear that 'strengthening the rule of law ... is critical in crisis affected countries as a key element of state and peace building', and is unequivocal about the umbilical link between securing the rule of law and development (echoing the High Level Declaration of the UNGA of 2012[62]):

> In September 2012, Member States in the General Assembly of the United Nations reaffirmed that the rule of law is critical for sustainable development. Indeed, the rule of law is an important factor in accelerating achievement of the MDGs and will be essential to the post-2015 Development Agenda as both an enabler and an outcome of development in its own right. It is now beyond question that improving safety for individuals and communities, and providing access to fair and well-functioning legal systems that adhere to international human rights standards, are necessary to promote economic investment, prevent violence and conflict, encourage inclusive growth and eradicate poverty. To these ends, governments, civil society groups and multilateral actors have in many cases increased the resources specifically devoted to the rule of law area.[63]

---

[58]Transforming our world: the 2030 Agenda for Sustainable Development, A/RES/70/1, 21 October 2015.

[59]http://www.undp.org/content/undp/en/home/ourwork/democratic-governance-and-peacebuilding/rule-of-law%2D%2Djustice-and-security.html—see video on this page: UNDP and the Rule of Law (last accessed 10 August 2019).

[60]http://www.undp.org/content/undp/en/home/ourwork/democratic-governance-and-peacebuilding/rule-of-law%2D%2Djustice-and-security.html—see video on this page: UNDP and the Rule of Law (last accessed 10 August 2019).

[61]http://www.undp.org/content/undp/en/home/ourwork/democratic-governance-and-peacebuilding/rule-of-law%2D%2Djustice-and-security.html—see video on this page: UNDP and the Rule of Law (last accessed 10 August 2019).

[62]See A/RES/67/1, para 7, 8, 24, 25, 30, 41. See section X above.

[63]UNDP (2014) Why, what and how to measure: a user's guide to measuring rule of law, justice and security programmes. http://www.undp.org/content/dam/undp/library/crisis%20prevention/UNDP_CPR_ROLMEGuide_August2014.pdf (last accessed on 10 August 2019).

Likewise, in the UNDP's strategic plan for 2014 to 2017, *Changing with the World*, rights regarding and security enhancing objectives are both evident. The plan features the goal of 'reinforcing the rule of law and citizen security' as a 'critical aspect of well-being'.[64] It notes that progress towards this important strategic goal will include both 'security sector reform', measures to 'ensure respect for citizen rights' and 'stronger civilian oversight' as well as 'faster progress in reducing gender-based violence', 'policing for reduction of crime and anti-social behaviours and attitudes'. All of these are said to 'improve prospects for success on citizen security'.[65] The UNDP's strategic goals are clearly now aligned with the ambitious violence reduction goals of the SDG's. No matter how well committed the UNDP may be to softer community enhancing and rights regarding forms of peacekeeping, the concrete goals of violence reduction also necessitate a commitment to shoring up the coercive structures of member States. This is simply unavoidable.

This shift in the securitization of the rule of law has not gone unnoticed. As Bergling, Wennerstrom, and Sannerholm argue: 'the character of international community responses to rule of law threats and challenges has a strong focus on security. Rule of law promotion taking place in UN and EU missions has undergone a 'securitisation' in how reforms are conceived and put into practice'.[66]

Indeed, some view the shift as deeply corrosive of the concept of the rule of law. As Humphreys argues:

> Where the rule of law comes to signify the coercive capacity of the state and its disciplinary structures and effects, rather than its constraints; when it signifies the rights of victim, rather than the accused; when it is measured in convictions and boots on the ground, rather than in the texture of the social fabric, something has surely been lost. When such an effect is promoted consistently, country-by-country at global level, through numerous channels, that loss begins to look irrevocable.[67]

We may well ask too whether this focus on peacekeeping is potentially just as corrosive of the notion of development which until now has clung to a softer community-based approach to security and rule of law intervention. But if this discrete approach to security is to be maintained, a greater honesty about the risk of securitization within the SDGs and SDG 16 needs to be sustained. What we see in SDG 16 is a goal of empowerment, but the licence it gives to the idea of crime control and the coercion of private violence isn't always about empowerment. It is also about social control.

---

[64]UNDP, Changing with the World, UNDP Strategic Plan: 2014-17, New York. Available at http://www.undp.org/content/undp/en/home/librarypage/corporate/Changing_with_the_World_UNDP_Strategic_Plan_2014_17.html (last accessed 17 August 2019).

[65]Changing with the World, pp. 30–31.

[66]Bergling et al. (2012), p. 99.

[67]Humphreys (2010), p. 174.

References

Bergling P, Wennerstrom P, Sannerholm R (2012) Rule of law and security sector reform: casual assumptions, unintended risks and the need for norms. Hague J Rule Law 4:98–119
Humphreys S (2010) Theatre of the rule of law: transnational legal intervention in theory and practice. Cambridge University Press
Lazarus L (2015) The right to security. In: Cruft et al (eds) The philosophical foundations of human rights. Oxford University Press, Oxford, pp 423–441

Liora Lazarus BA (UCT), LLB (LSE), DPhil (Oxon), is an Associate Professor in Law, and a Fellow of St. Anne's College, University of Oxford. Her primary research interests are in comparative human rights, security and human rights, comparative theory and comparative criminal justice. Born and raised in South Africa, she studied African Economic History at the University of Cape Town and Law at the London School of Economics and Political Science. From 1994 to 1995 she was a Fellow of the Max Planck Institute for Foreign and International Criminal Law, Freiburg, Germany. She came to Oxford in 1995 to write her doctorate at Balliol College, after which she went on to become a law fellow at St Anne's College. Dr. Lazarus's various academic publications are in the area of prisoners' rights, criminal justice and security and human rights. She has completed public reports for the UK Ministry of Justice on balancing public protection with human rights, and on the relationship between rights and responsibilities. She has acted as an advisor to the UK Stern Review on Rape Complaints, was asked to give evidence to the Parliamentary Modern Slavery Bill Committee, and was part of an International Committee of Experts to the Basque Working Group on Treatment of Political Detainees. Dr. Lazarus co-founded and is actively involved in the work of Oxford Pro Bono Pulico. She is also an Associate of the Oxford Human Rights Hub. Dr. Lazarus is an Academic Affiliate of the Bonavero Institute of Human Rights at Oxford and a Member of Oxford's Centre for Criminological Research. She sits on the Advisory Board of the Max-Planck Encyclopedia of Comparative Constitutional Law. Dr. Lazarus founded and now acts as the series editor of the Hart Studies in Security and Justice.

Climate Change, Sustainable Development, and Human Rights

Alan Boyle

Contents

Abstract Climate change is a global problem. Addressing it is one of the UN's Sustainable Development Goals, which the 2015 Paris Agreement is intended to implement. It will also affect the enjoyment of human rights in many ways, but its causes, its effects, and those responsible, are too numerous and too widely spread to respond usefully to individual human rights claims or to analysis by reference to particular human rights. The 2015 Paris Agreement is relevant to human rights law, not for what it says about human rights—which is next to nothing—but for what it says about the need to address the risk of climate change taking global temperatures above 1.5 °C. The UN special rapporteur is right in principle to argue that human rights law as a whole requires states to comply with expectations set out in Articles 2, 3 and 4 of the Agreement. UN human rights bodies need to act accordingly and hold states to account for what they have agreed in Paris.

A revised and updated version of this chapter was published in 67 *ICLQ* (2018) pp. 759–777.

A. Boyle (✉)
University of Edinburgh, Edinburgh Law School, Edinburgh, UK
e-mail: Alan.Boyle@ed.ac.uk

M. Kaltenborn et al. (eds.), *Sustainable Development Goals and Human Rights*, Interdisciplinary Studies in Human Rights 5,
https://doi.org/10.1007/978-3-030-30469-0_10

1 Introduction

Climate change is a threat to life on Earth as we know it. The Intergovernmental Panel on Climate Change ("IPCC") concludes that evidence of changes in the climate system is now unequivocal, with the atmosphere and oceans warming, glaciers and polar ice melting, sea levels rising, and greenhouse gas ("GHGs") concentrations increasing.[1] Atmospheric greenhouse gas concentrations now stand at levels unprecedented in the past 800,000 years. The oceans have mitigated climate change, with 30% of the anthropogenic CO_2 emissions being absorbed, making the oceans warmer and more acidic.[2] The Kyoto Protocol, adopted in 1997 as a first step towards stabilising the climate, has been ineffective in reducing the increased concentration of atmospheric GHGs. On present trends, even if the 2015 Paris Agreement is fully implemented by all parties, it may be difficult to keep the increase in global temperatures below 2 °C, let alone achieve the target of 1.5 °C.[3]

Unless we succeed in reducing and restraining the increase in global average temperatures the consequences for humanity will be increasingly serious, if not catastrophic. Already, the effects of climate change and the warming of the atmosphere and the oceans are starting to impact on water supplies, agriculture, forests, small islands, low-lying coastal areas, and the marine environment. They are causing or exacerbating desertification, extreme weather, flooding, loss of biodiversity and changes in the distribution and abundance of fish stocks. Various human rights are potentially affected: life, health, private life, property, the right to water, and the right to food and an adequate standard of living. Indigenous peoples, inhabitants of low lying islands and coastal areas, nomadic peoples, subsistence farmers and fishermen, may all be particularly vulnerable to the effects of climate change. So will the polar bear, the penguin and other wildlife species. But everyone will be vulnerable to climate change if attempts to control and mitigate its causes and effects do not succeed. This is a vision of unsustainable development over the long term. Thus the fundamental challenge posed by climate change is not so much to human rights here and now, but to future generations and humanity itself.

More than 30 years since it was first articulated by the Brundtland Report the concept of sustainable development remains almost infinitely malleable.[4] However defined, it is inherently complex and its implementation obliges governments to think in somewhat different terms from those to which they have been accustomed. Social, political and economic choices abound: what weight should be given to natural resource exploitation over nature protection, to industrial development over

---

[1]See generally, Intergovernmental Panel on Climate Change (2013). Hereafter "IPCC 2013".

[2]IPCC 2013, paras. 2.2, 3.3, 5.2, 6.2–3.

[3]IPCC, *Climate Change 2007: Synthesis Report* (Bangkok, 2007); UN Environment Programme ["UNEP"], *The Emissions Gap Report: Are the Copenhagen Accord pledges sufficient to limit global warming to 2° C or 1.5° C? A preliminary assessment* (Nairobi, 2010).

[4]See generally Redclift (1987); Jacobs (1991); Reid (1995); Moffatt (1995), esp. Ch 3; Goldin and Winters (1995); Dresner (2008); Neumayer (2013); Helm (2016).

air and water quality, to land-use development over conservation of forests and wetlands, to energy consumption over the risks of climate change? The Sustainable Development Goals adopted by the UN recognise that it is for each state to make these choices.[5] That may result in wide diversities of policy and interpretation, as different governments and international organisations pursue their own priorities and make their own value judgments, moderated only to some extent by international agreements on such matters as climate change and conservation of biological diversity.

Although "sustainable development" is used throughout the 1992 Rio Declaration on Environment and Development, and in Agenda 21, it was not until the 2002 World Summit on Sustainable Development that anything approaching a UN definition of the concept could be attempted. Three "interdependent and mutually reinforcing pillars of sustainable development" were identified in the Johannesburg Declaration—economic development, social development and environmental protection.[6] In 2015 the UN went further and adopted an ambitious set of "Sustainable Development Goals" ("SDGs").[7] Like Agenda 21 in 1992, the Agenda for Sustainable Development affirms the need for a "global partnership for sustainable development", and the SDGs are intended to promote implementation of the concept. This vision of sustainable development mainly focuses on economic and social development; reduction of poverty and hunger are the pre-eminent objectives, but the SDGs cover almost every aspect of international policy, including well-being for all ages, gender equality, safer cities, better sanitation, sustainable consumption, access to justice, and so on.

While the preamble to the Agenda for Sustainable Development fully recognises the seriousness of the environmental challenge—including threats to the climate, biodiversity, water resources, and the marine environment—only three of the seventeen sustainable development goals are of particular relevance to the environment. The first commits states to take "urgent action" to deal with climate change and its impacts.[8] At least in form they have now done so by adopting the 2015 Paris Agreement.[9] The second addresses conservation and sustainable use of the oceans and marine resources.[10] The third aims to conserve and sustainably manage terrestrial ecosystems such as forests and wetlands, combat desertification, and halt loss of biodiversity.[11] All three are of course inter-linked. None of this adds anything new to international environmental policy or law, but it does serve to reaffirm existing

[5] See below.

[6] UN, *Report of the World Summit on Sustainable Development* ["WSSD"], UN Doc. A/CONF.199/20 (2002) Resolution 1, para. 5.

[7] "Transforming Our World: the 2030 Agenda for Sustainable Development," UNGA Resolution 70/1 (2015).

[8] SDG 13.

[9] See below.

[10] SDG 14.

[11] SDG 15.

commitments within the context of a process whose outcomes the UN will review in due course.[12]

The SDGs have given the concept of sustainable development more concrete content, but they may also have under-estimated the seriousness of the environmental problems the world continues to generate on a global scale. This is the point made cogently by Helm, who observes that "It is not hard to make the case that the process of destruction [of natural capital] has gone too far, and that renewable natural capital as a whole is already well below the optimal level…The state of core ecosystem services, from climate to fresh water and soils, is already a serious cause for concern."[13]

The key argument in this critique of existing conceptions of sustainability is that natural capital is not infinitely substitutable by man-made capital, and that some natural capital must be preserved in order to meet the needs of future generations and prevent ultimate catastrophe.[14] Helm believes that international policy has been overly-focused on present generations, and that we need a stronger focus on the inheritance of future generations.[15] Climate change and its consequences for global biodiversity, the oceans, and water supplies illustrate his point: put simply this and other long term environmental problems have not been given nearly enough weight in the balance of environment and development. To address this omission Helm formulates an "aggregate natural capital rule" designed to ensure that future generations inherit a comparable stock of natural capital even if it is not the same stock.[16] He concludes that "no net loss would transform the environmental outlook."[17]

The "green economy" paradigm promoted by UN Environmental Programme ("UNEP") and the Organisation for Economic Cooperation and Development ("OECD") at the 2012 UN Conference on Sustainable Development ("UNCSD") was based on the same conclusion: it was intended to foster economic growth and development while ensuring that natural assets continue to provide the resources and environmental services on which human well-being now and in the future depends.[18] As the 2012 negotiations got underway, however, it soon became evident that the "green economy" would not turn into a new consensus, but would reopen unresolved questions concerning the very notion of sustainable development.[19] While some developed countries proposed the adoption of indicators to measure progress toward the green economy and the establishment of a capacity development scheme,

[12]UN General Assembly ["UNGA"] Resolution 70/1 (2015) paras. 72–91.

[13]Helm (2016), p. 63.

[14]Ibid, 54–62.

[15]Ibid, pp. 8, 40.

[16]Ibid, pp. 8, 40.

[17]Ibid, p. 8.

[18]UNEP, *Global Green New Deal* (Nairobi, 2009) and UNEP, *Towards a Green Economy: Pathways to Sustainable Development and Poverty Eradication* (Nairobi, 2011); OECD, *Towards Green Growth* (Paris, 2011); and World Bank, *Inclusive Green Growth* (Washington, 2012).

[19]Morgera and Savaresi (2012).

developing countries largely resisted these suggestions, arguing that the green economy should not turn into a "normative straightjacket."[20] The guidance on the green economy eventually included in the conference outcome document ("*The Future We Want*") thus represents only a minimalist common denominator between these largely opposing views.[21] It affirms that the green economy is but "one" of the tools available to achieve sustainable development, and should not be regarded as a "rigid set of rules."[22] All countries are encouraged to consider its implementation in a manner that "endeavours to drive sustained, inclusive and equitable economic growth and job creation."[23] Once again, the emphasis here is not on long-term sustainability of natural capital but on short term economic growth. This is the contested policy background against which we can now view the most recent agreement on climate change, the 2015 Paris Agreement.

2 The 2015 Paris Agreement

The Paris Agreement[24] sets out a new agenda for implementing the UN Framework Convention on Climate Change ("UNFCCC") and the UN sustainable development goals post-Kyoto.[25] Firstly it seeks to hold global temperature increases to "well below" 2 °C and if possible below 1.5°.[26] It achieves this objective principally by committing all states parties to "prepare, communicate and maintain successive nationally determined contributions" to reducing GHG emissions.[27] Secondly it seeks to enhance adaptation and climate resilience by promoting low carbon emissions development.[28] It achieves this objective mainly through provision for cooperation and capacity building and by reiterating the UNFCCC"s provisions on

[20] Earth Negotiations Bulletin, *Summary of the First PREPCOM for the UN Conference on Sustainable Development*: 17–19 May 2010, at p. 5.

[21] UNCSD, The Future We Want, UN Doc A/CONF.216/L.1 (2012). For analysis, see Morgera and Savaresi (2012).

[22] The Future We Want, para. 56.

[23] *Ibid,* para. 62.

[24] This is its full title. The Paris Agreement is in form, albeit not by name, a protocol to the 1992 UN Framework Convention on Climate Change ("UNFCCC"), with which it shares the same institutional features. It was adopted by decision of the parties to the UNFCCC and only parties to the UNFCCC may become parties to the Paris Agreement. It entered into force on 21 November 2016. See generally Bodansky (2016a).

[25] The commitments of developed state parties to reduce GHG emissions under the 1997 Kyoto Protocol expired in 2012 without having achieved any reduction in GHG emissions against the 1990 baseline.

[26] Article 2.

[27] Articles 3 and 4. See also Article 5 on conservation of carbon sinks (i.e. forests).

[28] Articles 2 and 7.

financing.[29] There is thus an implicit assumption in the Paris Agreement that sustainable development requires low carbon development and a cap on global temperature increases.

The Paris Agreement retains the controversial concept of common but differentiated responsibility on which the UN climate regime has until now been based,[30] but in a very different form. Unlike the Kyoto Protocol, all parties to Paris—not just the developed States parties—are expected to "prepare" some level of contribution to ensuring that GHG emissions peak as soon as possible and thereafter reduce rapidly so as to stabilise in the second half of the century.[31] The precise contribution for each party has not been agreed in advance, but will be determined unilaterally by each party in accordance with its capabilities. Developed States will still take the lead,[32] but developing States—notably China, the world's biggest GHG emitter, and India, the world's third biggest—are no longer exempt from making any emissions reductions, as they were under Kyoto. The understanding is that reductions are to increase progressively, insofar as each country's circumstances allow, "on the basis of equity, and in the context of sustainable development and efforts to eradicate poverty".[33] Paris does at least reflect reality: it recognises that climate change is not caused only by developed states and that it cannot meaningfully be addressed by simplistic ideas of historic responsibility. China and India have to play their part. This development is an important milestone in the evolution of the UNFCCC regime.

Will the Paris Agreement be more successful than Kyoto at reducing GHG emissions? On the positive side the UN climate regime for the first time has a clear and verifiable objective defined by reference to global temperatures. Equally importantly all the principal GHG emitters share a common commitment to reduce GHG emissions, even if their individual contributions will still vary and be nationally determined. On the negative side, based on commitments made so far, global temperatures will continue to rise well beyond 2° unless states progressively and significantly keep reducing their emissions.[34] The Agreement could work, or it could fail by a large margin, but those who want to influence the outcome can still do so. That includes the human rights community.

---

[29]Articles 6, 7, 9.

[30]Preamble, 3rd recital. See generally Rajamani (2006), especially Ch 6, and *id*, (2016).

[31]Article 4.

[32]Article 4(4).

[33]Articles 3 and 4(3).

[34]UNEP, *The Emissions Gap Report: Are the Copenhagen Accord pledges sufficient to limit global warming to 2° C or 1.5° C? A preliminary assessment* (Nairobi, 2010).

3 Human Rights and the Environment

Why should environmental protection be treated as a human rights issue? There are several possible answers.[35] Most obviously, and in contrast to the rest of international environmental law, a human rights perspective directly addresses environmental impacts on the life, health, private life, and property of individual humans rather than on other states or the global environment. It may serve to secure higher standards of environmental quality, based on the obligation of states to take measures to control pollution affecting health and private life. It helps to promote the rule of law in this context: governments become directly accountable for their failure to regulate and control environmental nuisances, including those caused by corporations, and for facilitating access to justice and enforcing environmental laws and judicial decisions. Lastly, the broadening of economic and social rights to embrace elements of the public interest in environmental protection has given new life to the idea that there is, or should be, in some form, a right to a healthy or decent environment.[36]

Human rights jurisprudence relating to the environment has developed significantly in the past 25 years since the Rio Conference, but none of this owes anything to UN human rights bodies. An early attempt to adopt a UN declaration on human rights and the environment terminated in 1994 when an ambitious but politically controversial draft failed to secure the backing of states.[37] The Office of the High Commissioner for Human Rights (OHCHR) returned to the problem in 2009, emphasizing that "While the universal human rights treaties do not refer to a specific right to a safe and healthy environment, the United Nations human rights treaty bodies all recognize the intrinsic link between the environment and the realization of a range of human rights, such as the right to life, to health, to food, to water, and to housing."[38] Three theoretical approaches were subsequently identified.[39] The first sees the environment as a "precondition to the enjoyment of human rights". The second views human rights as "tools to address environmental issues, both procedurally and substantively". The third integrates human rights and the environment under the concept of sustainable development. Finally, the report also identifies "the

---

[35]See Boyle and Anderson (1996), Merrills (2007), Anton and Shelton (2011) and Boer (2015).

[36]See UNHRC, *Preliminary Report of Independent Expert on human rights obligations relating to a safe, clean, healthy and sustainable environment*, A/HRC/22/43 (2012) ["UNHRC *Preliminary Rept* (2012)"]; Knox and Pejan (2017).

[37]ECOSOC, Draft Declaration of Principles on Human Rights and the Environment, in *Human Rights and the Environment*, Final Report (1994) UN Doc E/CN 4/Sub 2/1994/9. See Popovic (1996).

[38]OHCHR, *Report on Climate Change and Human Rights*, UN Doc A/HRC/10/61 (2009) para 18 ["OHCHR 2009 Rept"]. On the difficult political background see Limon (2017), Ch 11.

[39]OHCHR, *Analytical Study on the Relationship between Human Rights and the Environment*, UN Doc A/HRC/19/34 (2011), paras. 2, 6–9 ["OHCHR 2011 Rept"].

call from some quarters for the recognition of a human right to a healthy environment" and notes the alternative view that such a right in effect already exists.[40]

An independent expert-special rapporteur (Professor John Knox) was appointed by the UN Human Rights Commission (UNHRC) in 2012 to report on the substantive and procedural dimensions of human rights obligations relating to the enjoyment of a "safe, clean, healthy and sustainable environment" and the role of various institutional mechanisms in implementing those obligations.[41] His reports articulate a largely uncontroversial account of what human rights law has contributed to environmental protection at the national and international level. The Special Rapporteur noted that "no global agreement sets out an explicit right to a healthy (or satisfactory, safe or sustainable) environment. . .Nor have the later conferences on sustainable development in Johannesburg in 2002 and Rio de Janeiro in 2012 proclaimed a right to a healthy environment."[42] However, his final report in 2018 advocates recognition of a right to a "safe, clean, healthy and sustainable environment", derived largely from existing human rights law.[43]

At the same time there is value in setting down those elements of human rights law already reflected in the jurisprudence and in national law, and on which some degree of international consensus is achievable. The rapporteur has at least demonstrated that regional environmental practice in Europe, Latin America and Africa has global significance, including the procedural rights enshrined in the Arhus Convention, the obligation to assess environmental impacts and ensure a reasonable balance between economic development and environmental protection, the need to implement and enforce applicable environmental standards, and to protect vulnerable groups such as indigenous peoples.[44] But while the UNHRC now recognises the environmental dimensions of human rights law,[45] including its relevance for sustainable development, it remains to be seen whether UN treaty bodies follow suit in a meaningful way.[46]

Does existing human rights law have any role in the process of tackling global climate change elaborated by the Paris Agreement? The OHCHR believes that it does. Its current policy asserts that "it is critical to apply a human rights-based approach to guide global policies and measures designed to address climate

[40] *Ibid*, para. 12.

[41] UNHRC *Preliminary Rept* (2012); *Mapping Report*, UN Doc A/HRC/25/53 (2013) ["UNHRC *Mapping Rept* (2013)"]; *Compilation of Good Practices*, UN Doc A/HRC/28/61 (2015); *Note by Secretariat*, UN Doc A/HRC/31/53 (2015). See generally Knox and Pejan (2017), Ch 1.

[42] UNHRC *Preliminary Rept* (2012), para. 14.

[43] UNHRC, *Report of the Special Rapporteur on the issue of human rights obligations relating to the enjoyment of a safe, clean, healthy and sustainable environment,* UN Doc A/HRC/37/59 (2018).

[44] See UNHRC *Mapping Report* (2013).

[45] UNHRC Resolution 31/8 (2016).

[46] The current practice of the Committee on the Rights of the Child, the Committee on the Elimination of Discrimination against Women, and the Committee on Economic Social and Cultural Rights with respect to climate change is reviewed in CIEL, *States' Human Rights Obligations in the Context of Climate Change* (Washington, 2018).

change."[47] Its website identifies three "essential attributes" of a human rights approach:

- As policies and programmes are formulated, the main objective should be to fulfil human rights.
- The *rights-holders* and their entitlements must be identified as well as the corresponding *duty-bearers* and their obligations in order to find ways to strengthen the capacities of rights-holders to make their claims and of duty-bearers to meet their obligations.
- Principles and standards derived from international human rights law—especially the Universal Declaration of Human Rights and the core universal human rights treaties, should guide all policies and programming in all phases of the process.[48]

In its submission to the Paris Conference in 2015 the OHCHR set out ten considerations that should guide states in the actions they take to address climate change.[49] Inter alia these include mitigating climate change and preventing negative effects on human rights; ensuring accountability and effective remedy for human rights harms caused by climate change; and guaranteeing equality, non-discrimination, and meaningful and informed participation in decision-making. States should co-operate to ensure an equitable outcome that delivers "low-carbon, climate-resilient, and sustainable development, while also rapidly reducing green-house gas emissions." The policy asserts that "Only by integrating human rights in climate actions and policies and empowering people to participate in policy formulation can States promote sustainability and ensure the accountability of all duty-bearers for their actions. This, in turn, will promote consistency, policy coherence and the enjoyment of all human rights."[50]

The UN human rights community is a late arrival at the climate change ball—very late. But if climate change is the problem, is human rights the answer, or even part of the answer? And is the OHCHR's vision of that answer plausible or realisable? For several reasons, it is far from clear that it is.

First, the whole tenor of the OHCHR's submissions to the Paris negotiations is anthropocentric insofar as it focused only on the harmful impact of climate change on the rights of humans, rather than on the environment as such. This is precisely the approach which ecological theorists have opposed because they believe it is insufficiently comprehensive and inconsistent with ecological reality and biological diversity.[51] By looking at the problem in moral isolation from other species and

[47] See OHCHR website entry, "Human Rights and Climate Change: Overview", consulted 28/5/17, at http://www.ohchr.org/EN/Issues/HRAndClimateChange/Pages/HRClimateChangeIndex.aspx.

[48] Summarising the OHCHR website entry, previous fn.

[49] See "Understanding Human Rights and Climate Change", OHCHR's submission to the 21st Conference of Parties to the UNFCCC (27 November 2015) and the "Key Messages" reproduced on its website.

[50] "Key Messages", ibid.

[51] See Eckersley (1992) and Gillespie (1997).

the natural world we simply reinforce the assumption that the environment and its natural resources exist only for immediate human benefit and have no intrinsic worth in themselves. But as we saw earlier, we cannot afford to ignore the fundamental value of natural capital—the climate, biodiversity, ecosystems, the marine environment and so on—in sustaining life on Earth. The Paris Agreement recognises this point.[52] It is tempting to describe OHCHR's vision of human rights in this context as conceptually imperialist. A kinder view might be that it is simply myopic, for there is no inherent reason why a human rights perspective should not take a longer term-perspective that balances competing values. Indeed, human rights law already recognises that environmental protection is a legitimate aim of public policy and law that may in some cases constrain or limit the exercise of the right to possessions and property or other human rights.[53]

Second, beyond a dogmatic reiteration of the importance of human rights, no real attempt is made by the OHCHR to explore or explain in any detail how its emphasis on fulfilling human rights as "the main objective" is supposed to save the climate from excessive warming in decades to come. The submissions do not go into detail and, not surprisingly, those who drafted the Paris Agreement evidently did not share the OHCHR's conception of its own centrality to the problem. Thus, right from the start we have a relationship that is poorly thought out, and on which only a limited consensus exists.[54]

4 Human Rights in the Paris Agreement

The preamble to the Paris Agreement acknowledges that climate change is a common concern of humankind, and that *"Parties should, when taking action to address climate change, respect, promote and consider their respective obligations on human rights, the right to health, the rights of indigenous peoples, local communities, migrants, children, persons with disabilities and people in vulnerable situations and the right to development, as well as gender equality, empowerment of women and intergenerational equity"*.

[52]Preamble, para. 13, refers to "the importance of ensuring the integrity of all ecosystems, including oceans, and the protection on biodiversity. . . ."

[53]*Matos e Silva Lda v Portugal* (1996) IV European Court of Human Rights ["ECHR"]; *Jacobsson v Sweden* No 2 (1998) I ECHR; *Katte Klitsche and de la Grange v Italy* (1994) ECHR Sers A/293B; *Pine Valley Developments Ltd v Ireland* (1991) ECHR Sers A/222, paras. 57–9; *Katsoulis and Ors v Greece* [2004] ECHR 321; *Fredin v Sweden* (1991) ECHR Sers A/192, paras. 41–51. See also *Apirana Mahuika and Ors v New Zealand* (2000) CCPR Comm No 547/1992, in which the UN Human Rights Committee ["UNHR Ctte"] upheld the state's right to conserve and manage natural resources in the interests of future generations provided this did not amount to a denial of the applicant's rights.

[54]See Limon (2017), Ch 11.

Despite much effort to achieve more during the negotiations, this is all the agreement says on the subject of human rights.[55] It is even less specific or prescriptive than the OHCHR submissions. As Rajamani observes, "This recital carefully circumscribes the impact of an explicit reference to human rights in the Paris Agreement."[56] Human rights law is neither incorporated into the Paris Agreement by this wording, nor does it explicitly constitute a standard by which the adequacy of efforts taken by the parties to implement the objectives of the Paris Agreement might be judged.[57] Firstly, this paragraph is in the preamble, not the body of the treaty. It thus helps only to identify the object and purpose of the treaty and its context, not to impose obligations on the parties.[58] Second, it uses the word "should", not "shall", which implies a less than wholehearted endorsement of the relevance of the various rights referred to.[59] The phrase "respect, promote and consider" further minimises any sense that the Agreement is reiterating a commitment to "fulfil" or "protect" human rights, the terms normally associated with human rights commitments.[60] This choice of words was deliberate. At best it is little more than a recognition that, as the preamble says, states should take into account their human rights obligations when taking action to address climate change. To suggest that it brings about a "true incorporation of human rights into the Paris Agreement" is very wide of the mark.[61] It does nothing of the kind. This preamble is not a triumph for the human rights lawyers.

Then there is the list of rights referred to in the preamble. It is a curious catalogue. The right to health is obviously relevant to climate change. But there is no mention of the right to private life, or life, or property, nor any of the social and economic rights that are also relevant, except the right to development. The word "sustainable" is notably absent here, although referred to elsewhere. Indigenous peoples, children, women, persons with disabilities all have rights, and all will be vulnerable to climate change, but why their rights should be singled out for special mention is not obvious. Overall this looks more like a list of categories designed to satisfy special interest groups rather than a serious attempt to address the relationship between human rights law and climate change.[62] It is easy therefore to start from a sceptical position about Paris and human rights. Essentially the agreement says nothing meaningful on the subject. Does this matter? Yes it does.

---

[55]See Klein et al. (2017), pp. 108, 114–117; Rajamani (2010); Duyck (2015).

[56]Rajamani (2017).

[57]See Rajamani (2017). Contrast 1988 UN Convention Against Illicit Traffic in Narcotic Drugs, Article 14(2), which provides in part: "Each Party shall take appropriate measures to prevent illicit cultivation . . .The measures adopted shall respect fundamental human rights. . ."

[58]Gardiner (2008).

[59]On the difference between "should" and "shall" see Nordquist (1993), II, pp. xlv–xlvi.

[60]Rajamani (2017); Klein et al. (2017), p. 115. Contrast UNHRC Resolution 31/8 (2016), which calls on states to "respect, protect, and fulfil human rights obligations".

[61]See Klein et al. (2017), p. 115.

[62]A point confirmed by Klein et al. (2017), p. 116.

Climate change is already regarded in international law as a "common concern of humanity".[63] Thus it is an issue in respect of which all states have legitimate concerns. The UN is therefore right in principle to make the connection between human rights and climate change.[64] The OHCHR's first report on the subject noted, however, that "[w]hile climate change has obvious implications for the enjoyment of human rights, it is less obvious whether, and to what extent, such effects can be qualified as human rights violations in a strict legal sense".[65] That report goes on to observe the multiplicity of causes for environmental degradation and the difficulty of relating specific effects to historic emissions in any one state. But this is only partly true.

Governments obviously have a responsibility to protect their own citizens from the harmful effects of climate change, and this is where the rights of the most vulnerable groups are most relevant. The rights to life, health, water and food, and the right to respect for private life and property, entail more than a simple prohibition on government interference: governments have a positive duty to take appropriate action to secure these rights,[66] as we can see in the human rights case law on environmental harm, where the problems were mainly a failure to regulate the activities of corporations, or to legislate on environmental matters, or to enforce existing environmental law.[67] At this level it is worth emphasising that human rights obligations towards those most affected by climate change will at the very least require governments to take appropriate steps to mitigate the risk of harm within their own borders. If this is what the OHCHR had in mind then it is on safe ground, but it will have achieved very little.

However, in the climate change context, where the impacts are global, the key question is not whether GHG emitting states have to mitigate the harm to their own citizens, but whether they also have a responsibility to protect people in *other* states from the harmful impacts of those emissions on the global climate. The inhabitants of sinking islands in the South Seas may justifiably complain of human rights violations, but who is responsible? Those states like the United Kingdom and Germany whose historic emissions have unforeseeably caused the problem? Or those states like China and India whose current emissions are foreseeably making matters worse? Or those states like the United States or Canada which opted out of Kyoto and failed to take adequate measures to limit further emissions so as to stabilise global temperatures at 1990 levels? Or their own governments in the

[63]1992 UNFCCC, Preamble, 1st recital.

[64]UNHRC Resolution 10/4 (2009); id, *Rept of the Independent Expert/Special Rapporteur etc,* UN Doc A/HRC/31/52 (2016). See generally Humphreys (2009), Knox (2009), Boyle (2012) and Atapattu (2015).

[65]OHCHR 2009 Rept, para. 70.

[66]See UNHR Ctte, General Comment No 6 on Article 6 of the 1966 International Covenant on Civil and Political Rights, 16th Session, 1982.

[67]See e.g. *Lopez Ostra v. Spain* (1994) 20 EHRR 277; *Guerra v. Italy* (1998) 26 EHRR 357; *Fadeyeva v. Russia* [2005] ECHR 376; *Öneryildiz v. Turkey* [2004] ECHR 657; *Taskin v. Turkey* [2004] ECHR, paras. 113–9.

Association of Small Island States, which may have conceded far too much when ratifying the Kyoto Protocol or in subsequent climate negotiations? In OHCHR's report the short answer is all of the above, and more.[68]

The legal problem with this answer is that human rights treaties generally require a state party to secure the relevant rights and freedoms for everyone within its own territory or subject to its jurisdiction.[69] The extraterritorial application of human rights law has normally arisen only where the State exercises jurisdiction or control over persons or territory abroad.[70] These precedents bear little resemblance to transboundary pollution, where the State only exercises jurisdiction over harmful activities located within its own territory, not over those who are harmed beyond its borders. They are even further removed from climate change.[71] It is much harder to frame such a problem in terms of the jurisdiction or control over persons or territory required by the human rights case law. That is why the OHCHR's focus on "rights-holders" and "duty bearers" is misplaced and essentially incoherent in the context of climate change. It fails to capture adequately a global perspective on the problem.

But even if we accept that human rights law has transboundary application, it will still be hard to show that parties to the UNFCCC regime, including the major GHG emitters, have failed to strike the right balance between their own state's economic development and respect for human rights in other states[72] when they have either complied with or are exempt from GHG emissions reduction targets established by

---

[68]See OHCHR 2009 Rept, paras. 86 and 99, and Limon (2017).

[69]1966 ICCPR, Article 2, but see *Legal Consequences of the Construction of a Wall in the Occupied Palestinian Territory, Advisory Opinion*, ICJ Reports (2004), 136, para. 109, and contrast the 1966 ICESCR which has no such limitation. The 1969 American Convention on Human Rights, Article 1 and the 1950 European Convention on Human Rights, Article 1, make no reference to territory, but require parties to ensure to everyone "subject to" or "within" their jurisdiction the rights set out therein. See generally De Schutter (2010), pp. 142–179.

[70]*Legal Consequences of the Construction of a Wall in the Occupied Palestinian Territory, Advisory Opinion*, ICJ Reports (2004), 136, paras. 109, 112; *Application of the International Convention on the Elimination of All Forms of Racial Discrimination (Georgia v. Russian Federation), Provisional Measures Order*, ICJ Reports (2008), 386, para. 109; *Ecuador v. Colombia (Admissibility)* [2010] Inter American Commission on Human Rights (IACHR) Report No. 112/10, paras. 89–100; *Alejandre, Costa, de la Pena y Morales v. Republica de Cuba* [1999] IACHR Report No. 86/99, para.23; *Coard v. United States* [1999] IACHR Report 109/99, para. 37; *Al-Skeini v. United Kingdom* [2011] ECtHR, paras. 130–42; *Öcalan v. Turkey* [2005] 41 European Human Rights Reports (EHRR) 985, para. 91; *Ilascu v. Moldova and Russia* [2005] 40 EHRR 46, paras. 310–19, 376–94; *Issa et al. v Turkey* [2004] 41 EHRR 567, para. 71; *Cyprus v. Turkey* [2002] 35 EHRR 30, para. 78. See also General Comment No. 31 adopted by the UNHR Ctte, UN Doc. HRI/GEN/1/Rev.7, 192 at 194ff, para. 10.

[71]See the arguments in Boyle (2012), p. 613. In 2016 a German court dismissed a case brought by a Peruvian farmer against a German energy company on the basis that RWE had no duty to protect Peruvians from flooding caused by climate change. The case is currently on appeal: see https://business-humanrights.org/en/rwe-lawsuit-re-climate-change. The Philippines Commission on Human Rights is currently investigating whether foreign fossil fuel companies have violated the human rights of Filipinos who have endured climate-related disasters.

[72]*Taskin v Turkey*, ECHR (2004), para. 119; *Hatton v. UK*, ECHR (2003) 28, paras. 98, 122–9 (Grand Chamber).

Kyoto and agreed by the international community as a whole.[73] Inadequately controlled transboundary pollution is clearly a breach of general international law,[74] and may in limited circumstances also engage those human rights which potentially entail cross-border jurisdiction.[75] But, given the terms of the Kyoto Protocol, and the essentially voluntary character of key provisions of the Paris Agreement,[76] it is far from clear that inadequately controlled climate change violates any existing treaty obligations or general international law. In those circumstances the argument that it nevertheless violates the existing human rights obligations of states is not easy to make.

5 Is There Another View?

The UN independent expert/special rapporteur appointed to report on climate change and human rights, Professor John Knox, has observed that "whether or not climate change legally violates human rights norms is not the dispositive question."[77] He takes a more insightful approach than the one adopted by OHCHR. His report emphasises the global character of the threat climate change poses for the enjoyment of human rights and focuses on the need for global cooperation to tackle climate change effectively in order to avert serious harm to those rights. He argues that obligations to protect human rights in the context of internal environmental harm "can also inform the content of the duty of international cooperation when that duty pertains to a global environmental challenge such as climate change."[78] On this view of human rights law "All states have a duty to work together to address climate change, but the particular responsibilities necessary and appropriate for each State will depend in part upon its situation."[79]

Building on the 2015 Paris Agreement, Professor Knox reiterates the obligation of states to assess the climatic effects of activities within their jurisdiction; to control the activities of business and industry when these may cause climate change; to facilitate public participation in decision-making with respect to climate change; and

[73]Greenhouse gas emissions reduction targets under Kyoto apply only to Annex I developed state parties, not to developing countries, including China, India and Brazil.

[74]*Pulp Mills on the River Uruguay Case*, ICJ Reports (2010), paras. 101, 187.

[75]E.g. the right of equal access to justice, on which see 1998 Arhus Convention on Access to Information, Public Participation in Environmental Decision-making and Access to Justice, Article 3(9).

[76]On the question how far the specific provisions of the Paris Agreement create legally binding obligations see Bodansky (2016b). In particular he notes, ibid, at 150, that "The Paris Agreement does not require parties to implement their NDCs; instead it simply requires parties to pursue domestic mitigation measures, an obligation they already have under the UNFCCC."

[77]UN Doc A/HRC/31/52 (2016), para. 36.

[78]Ibid, para. 45.

[79]Ibid, para. 46.

to provide effective remedies for those adversely affected. All of this reflects existing human rights-based environmental jurisprudence.[80] His most important conclusion is that by itself the Paris Agreement will not prevent "disastrous consequences for human rights if states merely meet the commitments they have undertaken so far." It follows that "[f]rom a human rights perspective, then, it is necessary not only to implement the current intended contributions, but also to strengthen those contributions to meet the target set out in article 2 of the Paris Agreement."[81]

Knox's human rights perspective is considerably more focused and persuasive than the submissions made by the OHCHR to the Paris negotiations. Its utility is partly rhetorical, but if used effectively by UN human rights bodies to focus attention on how states parties respond (or fail to respond) to commitments made in the Paris Agreement it has potential juridical significance, comparable to the economic and social rights from which it draws inspiration. This represents a significant contribution to the debate on human rights and climate change. It may suggest that economic and social rights provide the best framework for that debate at the global level. Economic and social rights are generally concerned with encouraging governments to pursue policies which create conditions of life enabling individuals or peoples to develop to their full potential. They are programmatic, entailing progressive realization in accordance with available resources, but nevertheless requiring states to "ensure the satisfaction of, at the very least, minimum essential levels of each of the rights".[82]

The debate on climate change and human rights thus broadens out into a focus on environmental quality and sustainability. But despite the efforts of the UN human rights treaty organs to invest the 1966 International Covenant on Economic, Social and Cultural Rights ("ICESCR") with greater environmental relevance, it still falls short of giving environmental quality and sustainability, including the quality and sustainability of the global environment, recognition as a significant public interest.[83] Crucially, there is no right to a safe, clean, healthy or sustainable environment in the Covenant.[84] Lacking the status of a right means that the environment can be trumped by those values that have that status, including the right to economic

[80]Supra, n. 67.

[81]UN Doc A/HRC/31/52 (2016), paras. 72–84. See also Knox (2016).

[82]UN Committee on Economic Social and Cultural Rights ["UNCESCR"], General Comment No 3: The Nature of States' Parties Obligations (1990) (interpreting Article 2 of the Covenant). See Craven (1995), Ch 3.

[83]The UNCESCR has adopted various General Comments relevant to the environment and sustainable development, notably General Comments 14 and 15, which interpret Articles 11 and 12 of the ICESCR to include access to sufficient, safe, and affordable water for domestic uses and sanitation. See UNCESCR, *General Comment No. 14: The Right to the Highest Attainable Standard of Health*, UN Doc E/C.12/2000/4, 11 August 2000; *General Comment No. 15: The Right to Water*, UN Doc E/C.12/2002/11, 20 January 2003.

[84]Though there is a right to health and environmental hygiene (Article 12). Compare UNHRC Resolution 31/8 (2016), para. 5(a), which encourages States "To adopt an effective normative framework for the enjoyment of a safe, clean, healthy and sustainable environment."

development and natural resource exploitation.[85] This is an omission that needs to be addressed if the sustainability of the global environment as a public good is to receive the weight it deserves in the balance of economic, social and cultural rights.

The key question therefore is what values we think a covenant on economic and social rights should recognize in the modern world. Is protection of the global environment, including the global climate system, a sufficiently important public good to merit economic and social rights status comparable to economic development? The answer to this question surely has to be yes. The UN has repeatedly endorsed the promotion of sustainable development as the core principle of international environmental law and policy for all states.[86] This is reflected in the Paris Agreement's references to sustainable development.[87] Is it now time to ensure that it is also reflected in human rights law?[88]

The challenge posed by sustainable development is to ensure that environmental protection is fully integrated into economic policy. Acknowledging that the environment is part of this equation, both the 1992 Rio Declaration (Principle 3) and the 1993 Vienna Declaration on Human Rights (paragraph 11) emphasize that "[t]he right to development should be fulfilled so as to meet equitably the developmental and environmental needs of present and future generations". For political reasons—because the issue remains controversial—this is a perspective wholly absent from the OHCHR's vision of the role of human rights law in the Paris negotiations.[89] Yet, as a matter of law, the International Court of Justice (ICJ) has repeatedly referred to "the need to reconcile economic development with protection of the environment [which] is aptly expressed in the concept of sustainable development".[90] The essential point is that, while recognising that the right to pursue economic development is an attribute of a State's sovereignty over its own natural resources and territory, it cannot lawfully be exercised without regard for the detrimental impact on the sustainability of the global environment or on human rights in general. The Paris Agreement is important precisely because it provides a clearer yardstick by which to measure that detrimental impact than previous climate change agreements have done.

---

[85]Pursuant to Article 1 of both 1966 Covenants. See Merrills (2007), p. 666.

[86]*Report of the World Summit on Sustainable Development*, UN Doc. A/CONF.199/20, 26 August–4 September 2002, Resolution 1, para. 5.

[87]See Preamble and Articles 2, 4, 5, 6, 7, and 8.

[88]See generally Knox and Pejan (2017).

[89]See Limon (2017) and Rajamani (2017).

[90]*Pulp Mills on the River Uruguay Case*, ICJ Reports (2010) 14, para. 177; *Gabčíkovo-Nagymaros Project (Hungary/Slovakia),* ICJ Reports (1997) 7, para. 140; *Iron Rhine Case*, PCA (2005); Higgins (1999).

6 Conclusions

The Paris Agreement is so far the first, but hopefully not the last, attempt to promote the 'urgent action' to deal with climate change envisaged by SDG 13. This conclusion is obvious and uncontroversial. Beyond that, it is harder to draw clear conclusions about how far Paris has embraced a human rights approach to climate change. The Paris Agreement is relevant to human rights law, but not for what it says about human rights—which is next to nothing—but for what it says about the need to address the risk of climate change taking global temperatures above 1.5 °C. In order to help ensure that states parties meet this goal, and avoid the serious harm to sustainable development and the enjoyment of human rights that will otherwise result, the UN special rapporteur is right in principle to argue that human rights law as a whole requires states to comply with expectations set out in Articles 2, 3 and 4 of the Agreement. UN human rights bodies need to act accordingly and hold states to account for what they have agreed in Paris. Climate change is a global problem. It cannot easily be addressed by the simple process of giving specific human rights transboundary effect. It affects too many states and too much of humanity. Its causes, its effects, and those responsible, are too numerous and too widely spread to respond usefully to individual human rights claims or to analysis by reference to particular human rights.

The response of human rights law—if it is to have one—needs to be in global terms, treating the global environment and climate as the common concern of humanity, and climate change as a threat to human rights as a whole, but in particular to economic and social rights. In that context focusing on climate change within the corpus and institutional structures of economic, social and cultural rights makes sense, even if it means giving a broader interpretation to those rights or amending the 1966 ICESCR. The policies of individual states on energy use, reduction of greenhouse gas emissions, land use and deforestation could then be scrutinised and balanced against the evidence of their global impact on human rights and sustainable development. This is not a panacea for failure to make progress in implementing the Paris Agreement, but it would give the rights of humanity as a whole a voice that at present is scarcely heard. Whether the UN Human Rights Commission or the Office of the High Commissioner for Human Rights wish to travel down this road is another question, for politicians in Geneva to answer rather than international lawyers, but that is where they must go if they wish to say anything meaningful about climate change and human rights.

References

Books

Anton D, Shelton D (2011) Environmental protection and human rights. CUP, Cambridge

Atapattu S (2015) Human rights approaches to climate change: challenges and opportunities. Routledge, Abingdon

Boer B (ed) (2015) Environmental law dimensions of human rights. OUP, Oxford
Boyle A, Anderson M (eds) (1996) Human rights approaches to environmental protection. OUP, Oxford
Craven M (1995) The international covenant on economic, social and cultural rights. OUP, Oxford
De Schutter O (2010) International human rights law: cases, materials, commentary. CUP, Cambridge
Dresner S (2008) The principles of sustainability, 2nd edn. Routledge, London
Eckersley R (1992) Environmentalism and political theory. UCL Press, London
Gardiner R (2008) Treaty interpretation. OUP, Oxford
Gillespie A (1997) International environmental law, policy and ethics. OUP, Oxford
Goldin I, Winters L (eds) (1995) The economics of sustainable development. CUP, Cambridge
Helm D (2016) Natural capital: valuing the planet. Yale UP, London
Humphreys S (ed) (2009) Human rights and climate change. OUP, Oxford
Intergovernmental Panel on Climate Change (2013) Climate Change 2013: The Physical Science Basis: Working Group I Contribution to the Fifth Assessment Report of the Intergovernmental Panel on Climate Change. CUP, Cambridge
Jacobs M (1991) The green economy. Pluto Press, London
Klein D, Carazo M, Doelle M, Bulmer J, Higham A (eds) (2017) The Paris Agreement on climate change. OUP, Oxford
Knox J, Pejan R (eds) (2017) The human right to a healthy environment. CUP, Cambridge
Moffatt I (1995) Sustainable development: principles, analysis and policies. CRC Press, London
Neumayer E (2013) Weak versus strong sustainability, 4th edn. E Elgar, London
Nordquist M (ed) (1993) United Nations Convention on the Law of the Sea 1982: a commentary. Nijhoff, Dordrecht
Rajamani L (2006) Differential treatment in international environmental law. OUP, Oxford
Redclift M (1987) Sustainable development: exploring the contradictions. Methuen, London
Reid D (1995) Sustainable development. Earthscan, London

Book Chapters

Higgins R (1999) Natural resources in the case law of the international court. In: Boyle A, Freestone D (eds) International law and sustainable development. OUP, Oxford, p 87
Knox J (2016) Human rights principles and climate change. In: Carlarne C, Gray K, Tarasofsky R (eds) The Oxford handbook of international climate change law. OUP, Oxford, p 213
Limon M (2017) The politics of human rights, the environment, and climate change at the human rights council: towards a universal right to a healthy environment? In: Knox J, Pejan R (eds) The human right to a healthy environment. CUP, Cambridge, Ch 11
Merrills J (2007) Environmental rights. In: Bodansky, Brunnée, Hey (eds) The Oxford handbook of international environmental law. OUP, Oxford, p 663
Rajamani L (2017) Human rights in the climate change regime: from Rio to Paris and beyond. In: Knox, Pejan (eds) The human right to a healthy environment. CUP, Cambridge, Ch 13

Journal Articles

Bodansky D (2016a) The Paris climate change agreement: a new hope? Am J Int Law 110:306
Bodansky D (2016b) The legal character of the Paris Agreement. Rev Eur Community Int Environ Law 25:142
Boyle A (2012) Human rights and the environment: where next? Eur J Int Law 23:613
Duyck S (2015) The Paris climate agreement and the protection of human rights in a changing climate. Yearb Int Environ Law 26:3
Knox J (2009) Linking human rights and climate change at the UN. Harv Environ Law Rev 33:477
Morgera E, Savaresi A (2012) A conceptual and legal perspective on the green economy. Rev Eur Community Int Environ Law 22:14
Popovic N (1996) In pursuit of human rights: commentary on the draft declaration of principles on human rights and the environment. Columbia Hum Rights Law Rev 27:487
Rajamani L (2010) The increasing currency and relevance of rights-based perspectives in the international negotiations on climate change. J Environ Law 22:391
Rajamani L (2016) Ambition and differentiation in the 2015 Paris Agreement: interpretative possibilities and underlying politics. Int Comp Law Q 65:493

Alan Boyle is Emeritus Professor of Public International Law at the University of Edinburgh. He is a specialist in public international law, especially international environmental law, law of the sea, international law-making and international litigation. He has worked regularly as a consultant or as counsel for various governments, international organizations and NGOs on general international law, international environmental law and law of the sea. Inter alia he has appeared as counsel in six cases brought under the 1982 UN Convention on the Law of the Sea. He is one of the leading practitioners in that field and in international environmental litigation, where he has been counsel in many of the leading cases. Educated at Oxford University, he has also taught at the University of London (Queen Mary College); University of Texas Law School; William and Mary College Law School, Virginia; the University of Paris and LUISS in Rome. He is a barrister and practices occasionally in the International Court of Justice and other international tribunals, mainly in environmental and law of the sea cases.

Reflecting on the Right to Development from the Perspective of Global Environmental Change and the 2030 Agenda for Sustainable Development

Imme Scholz

Contents

Abstract The conceptual and legal relationship between human rights, human development and environmental protection is not a straightforward one. The 2030 Agenda for Sustainable Development and the Paris Climate Agreement adopted in 2015 link improvements in human development to human rights and to mitigating global changes in climate and the environment. The UN Declaration on the Right to Development (UNDRTD) adopted in 1986, however, does not include any explicit obligation to protect the natural environment, and to contribute to the provision of global environmental goods. The article explains how global environmental change is defined, how it is linked with human development and how it manifests itself. Then, the article takes a closer look at the UNDRTD and how it could be linked with global environmental change. Finally, the article proposes two concepts that could help to situate the UNDRTD within the challenges of the twenty-first century as exemplified in the 2030 Agenda. First, humanity should be introduced as a third category of right-holders (in addition to individuals and groups). This would include future generations more explicitly than now and put the relationships between species or life-forms as interdependent parts of the web of life into focus. Second, therefore, the rights of life forms should be established to transcend the conceptual boundaries of human rights and develop norms that govern the interdependencies between humans as well as plants and animals in the broadest sense.

I. Scholz (✉)
German Development Institute/Deutsches Institut für Entwicklungspolitik (DIE), Bonn, Germany
e-mail: imme.scholz@die-gdi.de

M. Kaltenborn et al. (eds.), *Sustainable Development Goals and Human Rights*, Interdisciplinary Studies in Human Rights 5,
https://doi.org/10.1007/978-3-030-30469-0_11

1 Introduction

The UN Declaration on the Right to Development (UNDRTD) adopted in 1986 and the 2030 Agenda for Sustainable Development adopted in 2015 share a universal concept of development that refers both to individual and collective dimensions of prosperity and thus includes the rights of future generations.[1] They thus offer a definition of the relationship between development and human rights that is very relevant for the twenty-first century. The core norm of the UNDRTD has been defined later as "the right of peoples and individuals to the constant improvement of their wellbeing and to a national and global enabling environment conducive to just, equitable, participatory and human-centred development respectful of all human rights".[2] Development thus refers to a people-centred and participatory process that aspires to achieve all human rights, it encompasses both individual and collective rights, and it binds states to promote human wellbeing within the borders of their own territory as well as beyond them.[3] It compels states to engage in international cooperation in order to remove obstacles for development and the achievement of human rights.

This is not only relevant for a world characterized by economic globalization and increasing interdependencies but also in times where the rationale for international cooperation has to be defended against nationalist trends in many countries on all continents. In particular, the consensus among Western countries, within the European Union and the G7, that rule-based international cooperation is beneficial for their prosperity and their objectives in international relations, seems to be crumbling, under the pressures of nationalist framings of increasing domestic economic inequalities and the rise of competing nations, notably China.

The linkage made to the first and second generations of human rights by the UNDRTD is echoed by the preamble to the Paris Agreement on Climate Change under the UNFCCC, which was adopted in 2015. It is the first time that a multilateral environmental agreement makes such a reference. The preamble acknowledges "that climate change is a common concern of humankind" and that "Parties should, when taking action to address climate change, respect, promote and consider their respective obligations on human rights". The Paris Agreement also includes "provisions to promote gender equality, and participation, sustainable development, and poverty eradication as side-benefits, or, more generally, as a context for climate action. Thus,

[1]UNDRTD Article 1 states in its first paragraph: "The right to development is an inalienable human right by virtue of which every human person and all peoples are entitled to participate in, contribute to, and enjoy economic, social, cultural and political development, in which all human rights and fundamental freedoms can be fully realized."

[2]This definition was elaborated in a broad consultative process by the High-Level Task Force on the Right to Development established in 2010 by the Human Rights Council, and which can be considered to being authoritative (quoted after Vandenbogaerde 2013, p. 19).

[3]de Feyter (2013) and Fukuda-Parr (2016).

the Paris Agreement contributes to the development of a political narrative justifying climate action by reference to human rights".[4]

The 2030 Agenda for Sustainable Development, the other landmark agreement of the UN in 2015, links human development rather closely with environmental protection. The preamble specifies that the states that adopted the Agenda "are determined to end poverty and hunger, in all their forms and dimensions, and to ensure that all human beings can fulfill their potential in dignity and equality and in a healthy environment". The negotiation of the 2030 Agenda was rooted in two distinct policy communities and processes: the development community was engaged in defining new objectives for poverty reduction and social development, building on the Millennium Development Goals that expired in 2015. The environmental community (together with a smaller part of the development community) focused on elaborating the Sustainable Development Goals[5] that had been agreed during the UN Conference on Sustainable Development 2012 in Rio de Janeiro. Both processes converged in the Agenda whose elaboration had thus been informed both by the principles of the Rio Declaration 1992 and by 'other instruments such as the Declaration on the Right to Development' and recognizes the need to build societies 'that are based on respect for human rights (including the right to development)'.[6] Like the UNDRTD, the Agenda emphasizes participation and the need for international cooperation in order to implement its set of 17 Sustainable Development Goals.[7] However, "the SDGs themselves and the attached targets do not represent a firm rights orientation".[8]

Both agreements adopted in 2015, if implemented, aim at improving human development inter alia by protecting it against threats associated with global changes in climate and the environment. In other words, they have the potential to redefine the relationship between human and economic development on the one hand, and the natural environment on the other. The 2030 Agenda specifies 17 goals to improve human development and to protect the planet's ecosystems. The Paris Agreement sets out to keep average global warming below 1.5 °C and to achieve global zero net emissions by the middle of the century.

It is difficult to link these concerns directly with the UNDRTD as the latter does not include any explicit mention of the natural environment. The second paragraph of Article 1 (see footnote 1) mentions the "full sovereignty" of peoples over their

---

[4]Mayer (2016), pp. 109–110.

[5]The SDGs were proposed by Colombia and Guatemala in the preparation of the Rio Conference 2012 in order to strengthen international consensus on cooperation for solving pressing and new global environmental problems, and for linking them systematically with improvements in human development. They explicitly referred to the success of the Millennium Development Goals in building support and guiding development assistance expenditure towards them, Scholz (2017).

[6]Arts and Tamo (2016), p. 222.

[7]Arts and Tamo (2016), p. 228.

[8]Arts and Tamo (2016), p. 247.

"natural wealth and resources",[9] and the definition of the core norm of 2010 mentioned above includes the term "enabling environment". It is debatable, however, whether these terms explicitly include the obligation to protect the natural environment, and to contribute to the provision of global environmental goods. In the 1990s and early 2000s, developing countries often rather considered these concerns as an "obstacle to development" as they subjected the economic use of natural wealth and resources to environmental provisions and thus made it more costly. In this line of argumentation, these provisions and costs were juxtaposed to poverty reduction. Both agreements adopted in 2015 reflect the insight that their implementation requires fundamental changes in development pathways and in patterns of production and consumption in order to accommodate environmental and social concerns. Therefore, the agreements are both universal and equally valid for rich and poor countries, and do not describe development as a process of catching up with current prosperity patterns of the "North". The UNDRTD is universal, too, but is not as explicit with regard to the relevance of environmental objectives for human development as the previous two.

Do the Paris Agreement and the 2030 Agenda require a reinterpretation or even a reformulation of the Declaration on the Right to Development? The objective of this article is to explain how the far-reaching impacts of global climate and environmental change and the action required for mitigating or avoiding it make—together with other challenges—a reinterpretation of the right to development necessary, in political discourse and in decision-making, in public debate and with regard to the expectations of future human wellbeing connected with it.

This article therefore does not refer to the legal debate on the quality of the UNDRTD as an independent substantial right but to its importance in political discourse, the meanings it provides for debates on the content of development policies and processes, and on the rights and duties of states in this respect. The mainstream of the legal literature considers the UNDRTD as soft law and Sengupta's assertion that "RTD has by now been established as an international human right"[10] is rather a minority statement. In political discourses and negotiations, however, RTD is an important reference, and negotiation results reflect this. The Rio Declaration of 1992 mentions the Right to Development in Principle 3 which calls for equitably meeting the developmental and environmental needs of present and future generations (de Feyter 2013). In the context of the UNFCCC, the right to development was translated into the principle of common but differentiated responsibilities and respective capabilities (CBDR-RC) which reveals a limited understanding of its universal approach. The CBDR-RC principle led to differentiated obligations: rich countries had to move forward and reduce their greenhouse gas emissions under the

---

[9]UNDRTD Article 1 second paragraph states: "The human right to development also implies the full realization of the right of peoples to self-determination, which includes, subject to the relevant provisions of both International Covenants on Human Rights, the exercise of their inalienable right to full sovereignty over all their natural wealth and resources".

[10]Sengupta (2006), p. 35.

Kyoto Protocol, while developing countries only had to report their emissions. Under the Paris Agreement, this principle was reaffirmed, after controversial negotiations, which allows differential treatment in the light of *different national circumstances*; in the articles of the Paris Agreement, in particular on mitigation, finance and transparency; and by introducing new and dynamic ways of differentiation through the new principles of progression and highest possible ambition.[11] The 2030 Agenda reflects many concerns that developing countries have rightly been emphasizing in the debate on the right to development, namely "inequalities in the international financial system, greater participation of developing countries in global decision-making on economic policy, and promoting a fairer trade regime".[12] Rich countries from the North have been less successful in anchoring their interpretation of the right to development in the agenda, which focuses more on the responsibilities of national governments in developing countries than on their own, namely to establish "suitable domestic conditions in developing countries such as good governance, democracy and responsible economic management".[13] Nevertheless, the reference in the 2030 Agenda to the UNDRTD helps to enhance participation and non-discrimination.[14] In addition, the 2030 Agenda and the UNDRTD equally strongly emphasize the responsibility of states for implementation, through national policy efforts and international cooperation.

Summarizing, the Right to Development appears to address issues of distribution and burden-sharing both within and between countries, as obstacles for enjoying the right to development, and to favour approaches that are in the interest of today's poor. Global climate change and worsening environmental trends, however, raise questions that are more fundamental because these changes and trends result from the cumulative environmental effects of immense improvements in global human welfare achieved in the twentieth century. The following section will explain how global environmental change is defined, how it is linked with human development and how it manifests itself. After that, the article will take a closer look at the UNDRTD, how it could be linked with global environmental change, and make some proposals for concepts that could help to situate the UNDRTD within the challenges of the twenty-first century as exemplified in the 2030 Agenda.

2 Global Environmental Change and Human Development

Global environmental change is an area of research that started to emerge in the 1960s and 1970s. Two topics were seminal: in the natural sciences, a growing interest in understanding the effects of rising anthropogenic emissions of greenhouse

[11] Voigt and Ferreira (2016).

[12] De Feyter (2013), p. 1.

[13] De Feyter (2013), p. 1.

[14] Arts and Tamo (2016).

gases into the atmosphere created space for the study of global warming and climate change and its interactions with industrialization and its infrastructures, land use and biodiversity. In economics, the 1972 report of the Club of Rome on the "limits of growth"[15] stirred a debate on the interactions between exponential economic and population growth and a finite supply of resources, the role of technological change therein and the dynamics that lead to either collapse or sustainability.[16]

In 2015, an international group of leading scientists from various disciplines published an article on nine "planetary boundaries" for human use of the Earth ecosystems and natural resources in a world characterized already by global environmental change.[17] Out of the comprehensive interactions of land, ocean, atmosphere and life that together provide the conditions for the existence of the human species on Earth, the authors identified nine processes and systems that regulate the stability and resilience of the Earth System and that are affected by human-induced changes to the environment. These include climate change, land-system change, biodiversity losses (genetic diversity under threat), changes in biogeochemical flows of nitrogen and phosphorus (that is to modern agriculture), ocean acidification, atmospheric aerosol loading, stratospheric ozone depletion and novel entities that pollute air, water and soils (for example toxic compounds, genetically modified organisms, and nanomaterials). "Two of these, climate change and biosphere integrity, are what the scientists call 'core boundaries'. Significantly altering either of these core boundaries would drive the Earth System into a new state", which includes the crossing of thresholds towards dangerous and irreversible change (http://www.anthropocene.info/planetary-boundaries.php).

The depth and certainty of scientific knowledge on how close we are to these thresholds varies: with regard to ocean acidification, ozone depletion and freshwater use we know that human use is still within the safe area—improving in the case of ozone depletion but dangerously close to the boundary in the case of ocean acidification. Global warming is increasingly affecting freshwater resources. Changes in the climate and the land system are in the zone of uncertainty while the latter is quite close to the boundary towards high risks. Genetic diversity and phosphorus and nitrogen flows are in the high-risk zone as the rate of species extinction has increased to unprecedented levels in human history during the last five decades, as has the use of synthetic fertilizers in agriculture. It has not been possible to assess yet where we are with regard to the novel entities and atmospheric aerosol loading (http://www.anthropocene.info/planetary-boundaries.php).

Defined as the result of interactions between natural and societal processes at global level, global environmental change can be characterized first as being a transboundary process whose impacts affect poor people and poor countries

[15]Meadows et al. (1972).

[16]For a summary of the reception of the report by economics over time, and how the original computer simulation and scenarios conform with measured changes since 1972 see the summary and literature in Wikipedia: https://en.wikipedia.org/wiki/The_Limits_to_Growth.

[17]Steffen et al. (2015).

disproportionately, due to their geographical location and their high socio-economic vulnerability (for the case of climate change, see IPCC Working Group II 2014).

Second, causes and impacts of global environmental change connect separate places over time and space, and the structure of causation is diffuse in itself. This is particularly true for global warming but also holds for the other processes that regulate the functioning of the Earth. Global warming today is the cumulative effect of greenhouse gas emissions of the last 200 years. These emissions occurred mainly in the centres of the industrial revolution, in Western Europe and the United States, but emission levels were high in the bloc of the former Soviet Union, too. In the last 20–30 years, absolute emissions grew exponentially in China as well as in India, Indonesia and other economically successful Asian countries. Impacts such as increased frequency and intensity of extreme weather events (droughts, floods, heat waves, storms), spread of vector-borne diseases (malaria, dengue) and rising sea levels (which will make coastal zones uninhabitable and lead to the disappearance of small island states) occur depending on geographies and climate, and not necessarily at places where emissions were generated. The generations who created the fossil-fuel economy and benefitted from it, and whose production and consumption patterns created global warming, will long be dead once impacts become apparent and the norm.

Emissions are caused mainly by energy use, and energy still overwhelmingly depends on the burning of fossil fuels. In the case of large utilities and energy-intensive economic sectors responsibility for postponing the switch to renewable energy technologies can be attributed. At the same time, the dependency of economic reproduction, of consumption and lifestyles on energy use and fossil-fueled bound technologies at the same time diffuses responsibilities and make change cumbersome. Environmental impacts are associated with specific technological trajectories that generate income, employment, and fulfill individual and societal needs in specific ways.[18] Moreover, increasing levels of income and prosperity in developing countries facilitate the reproduction of fossil-fuel based consumption and production patterns there, often at lower levels of energy efficiency.

The long causation chains between causes and impacts of global warming over time and space make it difficult to attribute legal responsibility and to address losses and damage, both for individual victims of climate change and for countries who suffer strong impacts without being large emitters.

Third, impacts of global environmental change do not only occur after long periods of time but they cannot be stopped easily, and sometimes they trigger irreversible processes of change with cascading effects.[19] Examples include the melting of the ice shields in Greenland and the Antarctica, the weakening of the Gulf Stream that makes temperatures in Europe more amenable than usual in higher latitudes, and mass extinction of species.

[18]Leach et al. (2010).

[19]World Bank 2012).

Fourth, the nine processes identified by Will Steffen and colleagues also interact with each other in complex ways. In most cases, linked impacts reinforce each other so that total impact is larger than the sum of all single impacts.[20]

Climate change is but one, albeit core, aspect of global environmental change. Since a few decades, environmental research is making efforts to understand the human impact on the earth system as a whole, by measuring the increasing human share in overall energy and material flows and the impacts this has on ecosystems and their functions.[21] As a whole, human impact has put the planet's ecosystems and their ability to sustain human life on earth under huge stress.[22] Human activity has changed half of the world's land surface, global energy and material flows exceed any natural flows, water resources, soils, forests and oceans are overexploited, and biodiversity is drastically reduced. This notion that the relationship between human society and nature has reached a new, unprecedented quality with far-reaching consequences for both led Nobel prize winner Paul Crutzen to define the present period since industrial revolution as a new geological epoch and to call it the Anthropocene.[23] The Anthropocene "marks the beginning of a new era of responsibility, as in terms of technology, humankind has by now advanced so far that it could unbalance the Earth system to an extent that would have dire consequences for human societies and ecosystems".[24]

Social environmental research shows that it has become increasingly difficult to empirically disentangle nature and society.[25] The processes of environmental overuse are produced and shaped by specific social practices that are embedded in and constitute economic and social subsystems (agriculture/food systems, cities, transport) that in turn are simultaneously linked to these environmental trends in various ways. Social practices vary with local historical, socio-economic and institutional conditions, and at the same time are influenced by global production and consumption networks to which they are connected. Therefore, there is no uniform way in how levels of human development and levels of environmental use are linked with each other.

The indeterminate relationship between prosperity and environmental damage becomes clear when comparing the levels of human development achieved by countries with their ecological footprints.[26] First, countries with a high level of human development have environmental footprints of nearly all sizes. No country, however, stays below the existing biocapacity per person. Second, the vast majority of countries whose environmental footprint is compatible with global sustainability as defined by the Global Footprint Network (GFN)—mostly from Africa—have low

[20]WBGU (2011, p. 45).

[21]This paragraph and the following three are taken from Scholz (2019) with minor changes.

[22]MA (2005).

[23]Crutzen (2002) and Steffen et al. (2007).

[24]WBGU (2011), p. 31.

[25]Leach et al. (2010).

[26]Wackernagel et al. (2017).

levels of human development. However, there are also poor countries with high environmental consumption. Third, there is no country that achieves high levels of human development within the environmental limits defined by the Global Footprint Network. As a trend, improvements in human development are linked with increasing environmental consumption.[27]

This trend, together with the large variations documented between countries, call for deeper analysis of the relationships between human society and nature, and how they are connected with and shaped by social inequalities, in order to explain both reinforcing linkages and pathways where decoupling between improved prosperity and environmental degradation can be observed. This is relevant for analytical and normative reasons and for better understanding current dominant development pathways, how they interact with social organization, with political and economic institutions, which in turn affect the ability to adopt changes required for making human prosperity sustainable and universal.

3 What Follows from This for the Right to Development? Human Development Within the Limits of the Earth's Ecosystems

Before I reflect upon what these considerations mean for the Right to Development, I first want to highlight some further aspects of the UNDRTD that are relevant for these reflections.

The Declaration on the Right to Development adopted by the UN General Assembly in 1986 does not define development as such but understands it to be a process for the full realization of all human rights. Development is declared to be an "inalienable human right" (Article 1) that depends on a comprehensive process that increases individual and collective wellbeing, and aspires to achieve social justice. States have the "primary responsibility for the creation of national and international conditions favourable to the realization of the right to development" (Article 3.1), i.e. to ensure that all persons living in their territory enjoy their rights as humans. They also have the "duty to co-operate with each other in ensuring development and eliminating obstacles to development" (Article 3.3), specifically by promoting a new international economic order. Article 4 establishes that states should implement adequate development policies, in their own territory and in support of developing countries, as a complement to their own efforts. Article 6 re-affirms the principle of international cooperation with a view to the respect of all human rights and specifically mentions that states "should take steps to eliminate obstacles to development resulting from failure to observe civil and political rights, as well as economic, social and cultural rights" (Article 6.3).

---

[27]Jackson (2017).

The reference to the new international economic order situates the Declaration on the Right to Development in the context of dependency theories and their understanding of the structural characteristics of post-colonial economies and societies as well as of international trade and production being an obstacle and not an opportunity for developing countries to benefit from international economic integration and growth. A development process that is guided by human rights and thus participatory in process and people-centred in its content therefore requires changes in these structures and in international political and economic relations.

Seen from this perspective, development seems to mean catching-up with industrialized countries, with regard to their material levels of wellbeing, their institutional structures and political orders. The right to development is derived from a critical view of supposedly universal principles that nevertheless allow for economic and political structures that systematically exclude developing countries.[28] Questions of environmental justice—negative social effects of environmental use, unequal distribution of pollution and other environmental bads to the detriment of the poor, and adverse distributive effects of environmental policies—do not figure in the UNDRTD. Neither does global environmental change.

After 1986, there were two significant additions made to the UNDRTD by two documents, a conceptual one and an institutional one:

- In 1993, the Vienna Declaration of the World Conference on Human Rights reaffirmed the Right to Development in a succinct and abbreviated version in its Article 10.[29] It then connected it with the concept of sustainable development (adopted in the final declaration of the UN Conference on Environment and Development in Rio de Janeiro in 1992) by stating that the right to development "should be fulfilled so as to meet equitably the developmental and environmental needs of present and future generations" (Article 11). This addition allows the DRTD to widen its normative horizon and include collective rights of future generations, juxtaposing them both to collective interests in fulfilling the needs of all in the present, and to strong individual or group-specific interests that may hamper meeting future needs. An important blank spot remains that has been filled by the 2030 Agenda: there is no acknowledgement of a "healthy

[28]For a deeper analysis of how the RTD is interpreted over time it makes sense to distinguish between its normative core and the specific calls for reforms of multilateral institutions and rules (such as the voting rights in the International Financial Institutions or multilateral trade and investment policy) that are seen as obstacles for the RTD.

[29]Article 10 of the Vienna Declaration 1993 reads: "The World Conference on Human Rights reaffirms the right to development, as established in the Declaration on the Right to Development, as a universal and inalienable right and an integral part of fundamental human rights. As stated in the Declaration on the Right to Development, the human person is the central subject of development. While development facilitates the enjoyment of all human rights, the lack of development may not be invoked to justify the abridgement of internationally recognized human rights. States should cooperate with each other in ensuring development and eliminating obstacles to development. The international community should promote an effective international cooperation for the realization of the right to development and the elimination of obstacles to development."

environment" as a prerequisite for development and that could be considered a (collective) human right, too.[30]

- The states' responsibility to act in relation to the right to development was specified in a document elaborated by a High-Level Task Force in 2010 that had been established by an Open-Ended Working Group under the Council of Human Rights. According to this report, "there exist three levels of States' responsibility (...): (i) States acting collectively in global and regional partnerships; (ii) States acting individually as they adopt and implement policies that affect persons not strictly within their jurisdiction; and (iii) States acting individually as they formulate national development policies and programmes affecting persons within their jurisdiction" (Vandenbogaerde 2013, p. 200). These three levels explain how states can act in order to ensure the individual and the collective dimension of the right to development[31]: through policies within their borders, by producing spillover effects across borders, and by explicit global or regional collective action. Collective rights and collective action are crucial for addressing the causes of global environmental change and for avoiding that rising levels of human prosperity increase environmental harm.

Making the Right to Development meaningful in a world increasingly marked by global environmental change requires concepts and norms that help to address the ways in which global environmental change interacts with specific ways for fulfilling human needs and achieving high levels of human development.

The historian Dipesh Chakrabarty has reflected about the question whether and how the Anthropocene requires new concepts when analyzing human prosperity, compared to the previous Earth age, the Holocene. His main conclusion is that a critique of capitalist globalization is insufficient for understanding human history in the age of the Anthropocene. First because the energy and material intensive industrial civilization is not limited to capitalist economies, and second because the Anthropocene "has brought into view certain other conditions for the existence of life in the human form that have no intrinsic connection to the logics of capitalist, nationalist, or socialist identities. They are connected rather to the history of life on this planet, the way different life-forms connect to one another, and the way the mass extinction of one species could spell danger for another. Without such a history of

---

[30]The Human Rights Council acknowledged the reinforcing relationship between the quality of the environment and the full enjoyment of human rights and appointed an independent expert on human rights and the environment in 2012, and later extended his mandate as special rapporteur in 2015. See http://www.ohchr.org/EN/Issues/Environment/SREnvironment/Pages/SRenvironmentIndex.aspx.

[31]The German strategy for sustainable development of 2016 for the implementation of the 2030 Agenda adopts a similar three-level approach: "Alongside measures with effects *in* Germany, there are also measures *by* Germany with a global impact. In addition, there is also the support of other countries in the form of bilateral cooperation (measures *with* Germany)", see Federal Government of Germany (2016), p. 3. Unfortunately, this definition omits international collective action.

life, the crisis of climate change has no human 'meaning'. For, as I have said before, it is not a crisis for the inorganic planet in any meaningful sense".[32]

Chakrabarty therefore proposes to introduce the category of human species for capturing the new dimension the Anthropocene adds to the future of humanity, and to cross-hatch the analysis of post-colonial capitalism with universal thinking in this sense. With this category it is possible to grasp the extent of the "shared catastrophe" all humans have fallen into once they became a geological agent, and to focus not only on internal conflicts within human society but to see it as part of the history of the web of life on the planet.

Is this a viable concept for making sense of the new quality of time and space covered by the relationship between human rights and the right to development in the Anthropocene? Democratic political systems, their institutions and procedures for making legitimate decisions are not adjusted to thinking in terms of the human species. Their concept of the collective has been elaborated and refined on the premise of national jurisdictions and laws. Moreover, democracies and their deliberations rely on the premise of an open future that individuals and groups can shape (and re-shape) according to their preferences.

Two fundamental dimensions of global environmental change are challenging for contemporary democracies: responsibility for decisions that have impacts over long times and for future generations (inter-generational justice) and beyond the national territory and across current generations (intra-generational justice). The cumulative effects of current production and consumption patterns over time reduce future opportunities for human prosperity—enjoying the freedom of an open society today will close the future for the generations to come (actually already for the current generation of children and teenagers) if no corrective action is taken against irreversible global environmental change. There are no practical rules for taking into consideration the rights of future generations into today's decisions. Extraterritorial responsibilities and the need for respective rules in the area of economic, social and cultural rights have been defined in the Maastricht Principles on Extraterritorial Obligations of States which recur to the UN Charter and human rights instruments, not to the UNDRTD. These principles are only slowly being accepted and translated into law and rules to be followed by non-State actors by some national governments. Others reject them. During the last four decades, economic globalization was advanced through deregulation and liberalization of national markets, and in this process, the notion of public interest to which collective public action is committed has weakened considerably. The less regulation in the public interest seemed necessary and legitimate for human prosperity at national level, the less it was possible to secure effective engagement in global public action for global public goods.

In an open, democratic society, these challenges can only be addressed if the destructive effects of human use of the environment are reflected upon collectively, and if the addressee of the rules is humanity as such (to reflect the threat to human

[32]Chakrabarty (2009), p. 217.

species, following Chakrabarty). This could mean to introduce humanity as a third right-holder in addition to individuals and peoples. From the perspective of nature-society relationships, however, this would also require norms about the relationship between humanity and other plant and animal species.[33] These challenges need to be taken up by dialogue, debate and decision-making, at local, national, regional and global levels of the executive, in legislatures and the judiciary, in other sectors of society, i.e. academia, the private sector, trade unions, and civil society organisations. Addressing the causes and the impacts of global environmental change on human prosperity, moreover, requires that international cooperation and collective public action in the interest of the global common good be strengthened and intensified.

Two proposals for qualifications to the right to development and, in consequence, as guidance for democratic decision-making, can be derived from this reasoning:

- Public/development policies in all countries have to respect the best knowledge on the limits of the earth's ecosystems and set their objectives and measures accordingly, in order to avoid trespassing local or global boundaries for safe human use of natural resources and sinks, and thus mitigate threats to the existence of humanity.
- Public/development policies in all countries shall not diminish the right to development/the full enjoyment of all human rights of others (intra- and intergenerational justice for humanity).

These qualifications and the reasoning and evidence that justify them need to be debated in order to adjust the international legal framework for human rights (and national law) not only to the challenges of economic globalization but also to those of global environmental change.

In 2013, there was a debate whether the UNDRTD should be strengthened by a legally binding framework convention[34] or whether this is not necessary as the Maastricht ETO Principles show.[35] Vandenbogaerde rightly says that the realization of human rights (and of development) requires individual agency, and that the duty of States is to enable individuals to carry out this agency.[36] Global environmental change, however, shows the limits of individual agency and the urgency of international collective action, as stated in the 2030 Agenda, in the twenty-first century. Maastricht ETO Principle 29 considers that: 'States must take deliberate, concrete and targeted steps, separately, and jointly through international cooperation, to create an international enabling environment conducive to the universal fulfilment of economic, social and cultural rights, including in matters relating to bilateral and multilateral trade, investment, taxation, finance, environmental protection, and

[33]In her most recent book, Donna Haraway reflects about how to increase humanity's empathy with other species, see Haraway (2016).

[34]De Feyter (2013).

[35]Vandenbogaerde (2013).

[36]Vandenbogaerde (2013), p. 200.

development cooperation'.[37] It is an advance that environmental protection is mentioned here, although at the same level of priority as financial and economic systemic issues. This ignores the risks of environmental change at species level.

Vandenboegarde goes on: "Both the right to development and the concepts of extraterritorial and transnational human rights obligations indicate that there is a fissure between today's human rights violations of a structural nature and the existing legal framework. Both share the idea that States and other powerful non-State actors have consequent obligations in filling this fissure, and both have to fight the reluctance or outright refusal of (mostly developed) States to acknowledge those obligations."[38]

It is clear that states (and non-state actors) do not only need to accept their extraterritorial responsibilities with a view to human rights and their foundation in the dignity of the individual, but also with regard to humans as a collective, the human species.

If humanity is introduced as a third category of right-holders (in addition to individuals and groups), then the norms that rule the relationships between species or life-forms as interdependent parts of the web of life come into focus. A third proposal consequently would be to transcend the conceptual boundaries of human rights and establish the rights of life forms, including humans as well as plants and animals in the broadest sense, and norms that govern their interdependencies.

Summarizing, the fundamental linkages between global environmental change and human prosperity, and the norms established in the 2030 Agenda, the Paris Agreement and the UNDRTD define five new tasks for states as duty bearers. First, states need to regard environmental protection and the provision of global environmental goods as fundamental and necessary areas of public policy and action for the full realization of all three generations of human rights (as defined in Articles 3, 4 and 6 of the UNDRTD). Second, this includes appropriate collective action at national, regional and global levels (legal frameworks, public policies and measures with regard to areas of public responsibility and with regard to respective guidance for non-state actors). Third, states have to ensure that they respect and fulfil extraterritorial obligations directly and indirectly towards people within and beyond their own jurisdiction and towards the global environment. Fourth, they have to respect the rights of future generations when making decisions today that reach out into the future. Finally, states should engage in first steps for developing a normative framework for the rights of all life-forms and the relationships between them.

[37]Quoted in Vandenbogaerde (2013), p. 206.

[38]Vandenbogaerde (2013), p. 208.

References

Arts K, Tamo A (2016) The right to development in international law: new momentum thirty years down the line? Neth Int Law Rev 63:221–249. https://doi.org/10.1007/s40802-016-0066-x

Chakrabarty D (2009) The climate of history: four theses. Crit Inq 35(2):197–222

Crutzen P (2002) Geology of mankind. Nature 415:23

De Feyter K (2013) Towards a framework convention on the right to development. Friedrich Ebert Stiftung, International Policy Analysis/Dialogue on Globalization, Berlin

Federal Government of Germany (2016) German Sustainable Development Strategy 2016. Federal Government, Berlin. https://www.bundesregierung.de/Content/DE/_Anlagen/2017/02/2017-02-27-nachhaltigkeit-neuauflage-engl.pdf?__blob=publicationFile&v=1

Fukuda-Parr S (2016) Human rights and politics in development. In: Goodhart M (ed) Human rights: politics and practice, 3rd edn. Oxford University Press, New York, pp 198–215

Haraway D (2016) Staying with the trouble: making kin in the Chthulucene. Duke University Press, Durham

IPCC WG II (Intergovernmental Panel on Climate Change Working Group II) (2014) Climate Change 2014: impacts, adaptation, and vulnerability. Summary for policymakers. IPCC, Geneva

Jackson T (2017) Prosperity without growth. Foundations for the economy of tomorrow, second expanded and revised edition. Routledge, Abingdon

Leach M, Scoones I, Stirling A (2010) Dynamic sustainabilities: technology, environment, social justice. Earthscan, London

MA (Millennium Ecosystem Assessment) (2005) Ecosystems and human well-being. Current state and trends. Island Press, Washington, DC

Mayer B (2016) Human rights in the Paris Agreement. Climate Law 6:109–117. https://doi.org/10.1163/18786561-00601007

Meadows DH, Meadows DL, Randers J, Behrens WW III (1972) The limits to growth. A report for the Club of Rome's Project on the predicament of mankind. Universe Books, New York. http://www.donellameadows.org/wp-content/userfiles/Limits-to-Growth-digital-scan-version.pdf. Retrieved 1 May 2018

Scholz I (2017) Herausforderung Sustainable Development Goals. In: Michelsen G (ed) Die Deutsche Nachhaltigkeitsstrategie: Wegweiser für eine Politik der Nachhaltigkeit. Hessische Landeszentrale für politische Bildung, Wiesbaden, pp 23–39

Scholz I (2019) The relevance of environmental research for development studies. In: Baud I, Basile E, Kontinen T, von Itter S (eds) Building development studies for the new millennium (EADI Global Development Series). Palgrave Macmillan, Basingstoke, pp 337–359. ISBN 978-3-030-04051-2

Sengupta A (2006) The human right to development. In: Marks SP, Andreassen BA (eds) Development as a human right: legal, political and economic dimensions. Harvard School of Public Health, Cambridge, MA

Steffen W, Crutzen PJ, McNeill JR (2007) The anthropocene: are humans now overwhelming the great forces of nature? Ambio 36(8):614–621

Steffen W et al (2015) Planetary boundaries: guiding human development on a changing planet. Science 347:1259855. https://doi.org/10.1126/science.1259855

Vandenbogaerde A (2013) The right to development in international human rights law: a call for its dissolution. Neth Q Hum Rights 31(2):187–209

Voigt C, Ferreira F (2016) 'Dynamic differentiation': the principles of CBDR-RC, progression and highest possible ambition in the Paris Agreement. Transnatl Environ Law 5(2):285–303. https://doi.org/10.1017/S2047102516000212

Wackernagel M, Hanscom L, Lin D (2017) Making the Sustainable Development Goals consistent with sustainability. Front Energy Res 5, Article 18. https://doi.org/10.3389/fenrg.2017.00018

WBGU (Wissenschaftlicher Beirat Globale Umweltveränderungen – German Advisory Council on Global Change) (2011) World in transition: a social contract for sustainability, Flagship Report 2011. WBGU, Berlin
World Bank (2012) Turn down the heat – why a 4°C warmer world must be avoided. World Bank, Washington DC

Imme Scholz is a sociologist and Acting Director of the German Development Institute (Deutsches Institut für Entwicklungspolitik DIE) since 2009. Before, she headed the department on environmental policy and natural resource management at DIE, and introduced work related to climate change into the institute's research and policy advisory. She is a member of the German Council for Sustainable Development since 2013. In 2015 she joined a CCICED task force on greening China's South-South Cooperation. She has published extensively on several topics at the interface of environment and development, including the 2030 Agenda and the SDGs, adaptation to climate change, and sustainable forest use. In the 1980s, she studied sociology at the Freie Universität Berlin where she did her doctorate in 1999 after joining DIE in 1992. Between 1999 and 2002, she was on leave and worked for German development cooperation as an environmental policy advisor in the Brazilian Amazon.

The Role of Public and Private Actors and Means in Implementing the SDGs: Reclaiming the Public Policy Space for Sustainable Development and Human Rights

Jens Martens

Contents

Abstract In the 2030 Agenda governments committed to a revitalized Global Partnership between States and declared that public finance has to play a vital role in achieving the Sustainable Development Goals (SDGs). But in recent decades, the combination of neoliberal ideology, corporate lobbying, business-friendly fiscal policies, tax avoidance and tax evasion has led to a massive weakening of the public sector and its ability to provide essential goods and services and to fulfill its human rights obligations. The same corporate strategies and fiscal and regulatory policies that led to this weakening have enabled an unprecedented accumulation of individual wealth and increasing market concentration. The proponents of privatization and public-private partnerships (PPPs) use these trends to present the private sector as the most efficient way to provide the necessary means for implementing the SDGs. But many studies and experiences by affected communities have shown that privatization and PPPs involve disproportionate risks and costs for the public sector and can

This article is based on the findings of the international civil society report Spotlight on Sustainable Development 2017. An earlier version was published as introductory chapter of this report on behalf of the Civil Society Reflection Group on the 2030 Agenda for Sustainable Development.

J. Martens (✉)
Global Policy Forum, Bonn, Germany
e-mail: jensmartens@globalpolicy.org

M. Kaltenborn et al. (eds.), *Sustainable Development Goals and Human Rights*, Interdisciplinary Studies in Human Rights 5,
https://doi.org/10.1007/978-3-030-30469-0_12

even exacerbate inequalities, decrease equitable access to essential services and jeopardize the fulfilment of human rights. Therefore, it is high time to counter these trends, reclaim public policy space and take bold measures to strengthen public finance, rethink PPPs and weaken the grip of corporate power on people's lives.

1 Re-defining the Global Partnership Agenda

When governments negotiated the 2030 Agenda in 2015 there were hard fights about the nature of a global partnership. While the G77 and its members from the global South emphasized the need for a revitalized global partnership among governments, the USA, the EU and their partners from the global North pushed for all kinds of partnership*s* between public and private actors to implement the Agenda and its goals. The latter followed the line of reasoning of the High-Level Panel of Eminent Persons on the post-2015 Development Agenda that stated in its final report in May 2013:

> We live in an age when global problems can best be solved by thousands, even millions, of people working together. These partnerships can guide the way to meeting targets and ensuring that programmes are effective on the ground. [. . .] These partnerships are powerful because each partner comes to the table with direct knowledge and strong evidence, based on thorough research. This enables them to innovate, to advocate convincingly for good policies, and thus to secure funding.[1]

In the context of the 2030 Agenda, the difference between partnership and partnership*s* is not just semantic sophistry but reflects two fundamentally different views of the role of the State: on the one hand as duty-bearer, particularly with respect to human rights, and as central provider of public goods and services, on the other hand as moderator and facilitator of actions of various public and private 'stakeholders'.

At the end of negotiations on the 2030 Agenda, governments agreed on a clearly graduated compromise: they fully committed to a revitalized Global Partnership at the governmental level and declared that public finance "will play a vital role in providing essential services and public goods and in catalysing other sources of finance."[2] But they also acknowledged the role of the "diverse private sector, ranging from micro-enterprises to cooperatives to multinationals, and that of civil society organizations and philanthropic organizations in the implementation of the new Agenda."[3]

In SDG 17 on means of implementation, governments included two targets under the subheading "Multi-stakeholder partnerships", but even there they first committed

[1]High-Level Panel of Eminent Persons on the post-2015 Development Agenda (2013), p. 22.
[2]UN (2015b), para. 41.
[3]Ibid.

to enhance the Global Partnership for Sustainable Development, only "complemented by multi-stakeholder partnerships" (target 17.16) and qualified the relevance of public-private partnerships by embedding them between public and civil society partnerships (target 17.17).

The embrace of the private sector and public-private partnerships became more visible in the outcome document of the Third International Conference on Financing for Development from July 2015, the *Addis Ababa Action Agenda* (AAAA).[4] This de facto funding programme for the SDGs devotes a separate chapter to the important role of private business and finance, and it contains 11 paragraphs that promote, welcome or encourage the use of multi-stakeholder or public-private partnerships.[5]

The trend towards partnerships with the private sector is based on a simple assumption: global problems are too big and the public sector is too weak to solve them alone.

2 Weakening the State: A Vicious Circle

The trend towards privatization and the promotion of public-private partnerships (PPPs) of various kinds are not at all new. The world faced a first wave of deregulation and privatization in the 1980s and 1990s, promoted by neoliberal policies of Western governments, advanced by the transition from centrally planned to market economies in Eastern Europe and the former Soviet Union, and imposed by Structural Adjustment Programmes of IMF and World Bank in highly indebted countries of the global South.

In the aftermath of the global financial crisis 2007–2008 the discourse around privatization and PPPs has gained new momentum, particularly shaped by corporate think tanks and international financial institutions (IFIs). At a time when governments seem unable and unwilling to resolve pressing challenges, private actors are positioning themselves as an alternative solution, more flexible, efficient and un-bureaucratic than governments.

A telling example of this strategy is the report of the World Economic Forum (WEF) on the future of global governance, "Global Redesign".[6] The report postulates that a globalized world is best managed by a coalition of multinational corporations, governments (including through the UN system) and select civil society organizations (CSOs). It argues that governments no longer are "the overwhelmingly dominant actors on the world stage"[7] and that "the time has come for a new stakeholder paradigm of international governance".[8] The World Economic

[4]UN (2015a).

[5]Ibid., paras. 10, 42, 46, 48, 49, 76, 77, 115, 117, 120 and 123.

[6]World Economic Forum (2010).

[7]Ibid., p. 8.

[8]Ibid., p. 9.

Forum vision includes a "public-private" UN, in which certain specialized agencies would operate under joint State and non-State governance systems, such as the Food and Agriculture Organization (FAO) through a "Global Food, Agriculture and Nutrition Redesign Initiative".[9] This model also assumes that some issues would be taken off the agenda of the UN system to be addressed by "plurilateral, often multi-stakeholder, coalitions of the willing and able".[10]

The IFIs, led by the World Bank, argued in a similar way in the discussions about the 2030 Agenda and the implementation of the SDGs. They called for a "paradigm shift on how development will be financed [. . .] to unlock the resources needed to achieve the SDGs."[11] In their view, the global community needs to move the discussion from "billions" in ODA to "trillions" in investments of all kinds, to meet the investment needs of the SDGs. While they admit that the majority of development spending happens at the national level in the form of public resources, they stress that the largest potential for additional funds is from private sector business, finance and investment. "This is the trajectory from billions to trillions, which each country and the global community must support together to finance and achieve the transformative vision of the SDGs."[12]

But why is it apparently a matter of fact that the public sector is too weak to meet the challenges of the 2030 Agenda? Why are public coffers empty? In fact, the lack of capacity and financial resources is not an inevitable phenomenon but has been caused by deliberate political decisions. To give just one example, over the past three decades corporate income tax rates have declined in both countries of the global North and South by 15–20%.[13] Hundreds of billions of US dollars are lost every year through corporate tax incentives and various forms of tax avoidance. Through their business-friendly fiscal policies and the lack of effective global tax cooperation, governments have weakened their revenue base substantially. This has been driven not least by corporate lobbying. A recent analysis by Oxfam America estimates that between 2009 and 2015, the USA's 50 largest companies spent approximately US$ 2.5 billion on lobbying, with approximately US$ 352 million lobbying on tax issues. In the same period, they received over US$ 423 billion in tax breaks.[14]

Widespread tax evasion and avoidance by transnational corporations and wealthy individuals make things even worse. It further decreases public revenues and exacerbates inequalities, as tax evasion seems to rise sharply with wealth. According to recent estimates by researchers in Norway, Sweden and Denmark, on average about 3% of personal taxes are evaded in Scandinavia, but this figure rises to about 30% in the top 0.01% of the wealth distribution, a group that includes households

[9]Ibid., p. 367.

[10]Ibid., p. 8.

[11]World Bank et al. (2015), p. 2.

[12]Ibid., p. 1.

[13]Crivelli et al. (2015).

[14]Oxfam America (2017), p. 2.

with more than US$ 40 million in net wealth.[15] The authors conclude: "Taking tax evasion into account increases the rise in inequality."[16]

What we see is a vicious circle of weakening the State: the combination of neoliberal ideology, corporate lobbying, business-friendly fiscal policies, tax avoidance and tax evasion has led to the massive weakening of the public sector and its ability to provide essential goods and services, as described in the report Spotlight on Sustainable Development 2017, inter alia, in its analyses on food security and sustainable agriculture (SDG 2), health (SDG 3), education (SDG 4), water (SDG 6), transport and housing (SDG 11).[17] These failures have been used by the proponents of privatization and PPPs to present the private sector as the better alternative and to demand its further strengthening. This in turn further weakened the public sector—and so on....

In parallel, the same corporate strategies and fiscal and regulatory policies that led to the weakening of the public sector enabled an unprecedented accumulation of individual wealth and increasing market concentration, often at the expense of small and medium-sized enterprises.

3 Concentrated Power

The globalization of the world economy and the waves of deregulation and privatization have facilitated the emergence and increased the power of large transnational corporations (TNCs) and financial conglomerates. Companies with activities in dozens of countries and billion-dollar turnovers have acquired both great influence on the global economic system and significant political clout.

According to various statistics of the largest national economies, transnational corporations, banks and asset management firms, among the 50 largest global economic entities are more private corporations than countries.[18] The assets under management by the world's largest asset management company BlackRock are US$ 5.12 trillion (end of 2016),[19] thus higher than the GDP of Japan or Germany.

Increasing market concentration has put greater power in the hands of a small number of corporations. An investigation of the relationships between 43,000 transnational corporations has identified a small group of companies, mainly in the financial industry, with disproportionate power over the global economy. According to the study by the Swiss Federal Institute of Technology in Zurich, "transnational corporations form a giant bow-tie structure and [...] a large portion of control flows

[15]Alstadsæter et al. (2017), p. 1.

[16]Ibid.

[17]Civil Society Reflection Group on the 2030 Agenda for Sustainable Development (2017).

[18]www.willistowerswatson.com/en/insights/2016/10/The-worlds-500-largest-asset-managers-year-end-2015.

[19]www.blackrock.com/de/privatanleger/uber-blackrock.

to a small tightly-knit core of financial institutions."[20] At the centre of the bow tie, a core of 147 companies control 40% of the network's wealth, while just 737 companies control 80%.

Large institutional investors such as pension funds, insurance funds and sovereign wealth funds are also the drivers of a new generation of PPPs in infrastructure, forcing governments to offer 'bankable' projects that meet the needs of these investors rather than the needs and rights of the affected population.[21]

Particularly alarming for the implementation of SDG 2 on food security and sustainable agriculture are the announced mega-mergers in the food and agriculture sector, especially the acquisition of Syngenta by China National Chemical Corporation (ChemChina), the merger of Dow Chemical and DuPont and the takeover of Monsanto by Bayer. If all of these mergers are allowed, the new corporate giants will together control at least 60% of global commercial seed sales and 71% of global pesticide sales.[22]

The growth and concentration of corporate power also includes private military and security companies (PMSCs). A 2011 study estimated the number of employees in this sector to be between 19.5 and 22.5 million, a number which exceeds the number of police officers worldwide.[23] The growth of this sector directly affects the implementation of SDG 16, as PMSCs can play a critical role in enabling violent conflicts by outsourcing political, economic and human costs.

4 Devastating Impacts

Privatization, PPPs and the rise of corporate power affect all areas and goals of the 2030 Agenda. One obvious example is the mushrooming of private, fee-charging, profit-making schools in Africa and Asia, with the particular case of Bridge International Academies, which operates 500 nursery and primary schools in Kenya, Uganda, Nigeria, Liberia and India.[24]

Detrimental corporate influence occurs in the energy sector with the still dominant role of coal and fossil fuel industries, undermining effective measures against climate change and the transformation towards sustainable energy systems. The extractive industries play a similar role, particularly with the rush to mine in the deep sea representing its newest frontier and perhaps the biggest threat to the world's oceans. Biodiversity and terrestrial ecosystems are equally threatened by the commodification of the values and 'services' provided by these industries, and by market-based conservation mechanisms. They risk marginalizing the actors that

[20]Vitali et al. (2011).

[21]Boys (2017), p. 93f.

[22]ETC Group (2017).

[23]Florquin (2011).

[24]Wulf (2017), p. 57.

play a central role in biodiversity conservation: indigenous peoples and local communities.

Studies by scholars, CSOs and trade unions like Public Services International (PSI) have shown that the privatization of public infrastructure and services and various forms of PPPs involve disproportionate risks for the affected people and costs for the public sector. They can even exacerbate inequalities, decrease equitable access to essential services, and thus jeopardize the fulfilment of human rights, particularly the rights of women.

Even evaluations done by the World Bank, the IMF and the European Investment Bank (EIB)—the organizations normally promoting PPPs—have found many cases where PPPs did not yield the expected outcomes.[25] Some of the findings of various studies on the risks and costs of PPPs can be summarized as follows:

- only very few countries have sufficient capacity to implement infrastructure PPPs;
- the cost of financing is higher for PPPs than for public sector works, as governments usually borrow at a lower rate than the private sector;
- potential short-term fiscal profits from large-scale PPPs are not always sufficient to offset the long-term additional costs arising from contract renegotiations;
- government liabilities for PPPs appear 'off-budget', so governments have the illusion that they have more fiscal space than they actually do.

Addressing the role of the G20 in a recent paper on infrastructure investment and PPPs, Nancy Alexander of the Heinrich Böll Foundation summarizes:

> The scale of the infrastructure and PPP initiative championed by the G20's national and multilateral banks could privatize gains and socialize losses on a massive scale. The G20 should take steps to ensure that this scenario does not unfold.[26]

5 Counter-Movements and Breaking Ranks

Responding to the experiences and testimonies from the ground about the devastating impacts of privatization and PPPs, counter-movements emerged in many parts of the world. Over the past 15 years there has been a significant rise in the number of communities that have taken privatized services back into public hands—a phenomenon called "remunicipalization".[27] Remunicipalization refers particularly to the return of water supply and sanitation services to public service delivery. Between March 2000 and March 2015 researchers documented 235 cases of water remunicipalization in 37 countries, affecting more than 100 million people.

[25]See references e.g., in Jomo et al. (2016) and Alexander (2016).

[26]https://us.boell.org/2016/12/15/infrastructure-investment-and-public-private-partnerships.

[27]Kishimoto (2016).

Furthermore, some pioneering companies are already on the path towards—at least environmentally—sustainable development solutions, for instance in the area of renewable energies. The private sector is in no way a monolithic bloc. Firms in the social and solidarity economy, social impact investors and small and medium-sized businesses are already making a positive difference, challenging the proponents of global techno-fix solutions and the dinosaurs of the fossil fuel lobby.

Even the firm opposition to international corporate regulation in the field of business and human rights by those pretending to represent business interests is showing cracks. A survey by The Economist Intelligence Unit revealed that a significant proportion of business representatives are now in favour of an international legal instrument to regulate corporate activities. The report concludes that:

> (…) although the reaction by most businesses has been negative, questioning not only the desirability but the efficacy and feasibility of such an instrument, 20% of respondents to our survey said that a binding international treaty would help them with their responsibilities to respect human rights.[28]

6 What Has to Be Done?

To be sure, the business sector certainly has an important role to play in the implementation process of the 2030 Agenda, as sustainable development will require large-scale changes in business practices. However, acknowledging corporations' role should not mean promoting the accumulation of wealth and economic power, giving them undue influence on policy-making and ignoring their responsibility in creating and exacerbating many of the problems that the 2030 Agenda is supposed to tackle.

Instead of further promoting the misleading discourse of 'multi-stakeholderism' and partnerships between inherently unequal partners a fundamental change of course is necessary. In order to achieve the SDGs and to turn the vision of the transformation of our world, as proclaimed in the title of the 2030 Agenda, into reality, we have to reclaim the public (policy) space. This includes, inter alia, the following steps:

Strengthening Public Finance at All Levels

Widening public policy space requires, among other things, the necessary adjustments in fiscal policies. In other words, governments have to formulate Sustainable Development Budgets in order to implement the Sustainable Development Goals. They can generally approach the issue from both the revenue (tax policy) and the expenditure (budget policy) angle. They can pursue proactive tax policies to achieve

[28] The Economist Intelligence Unit (2015), p. 23.

environmental and social policy goals and simultaneously fulfill their human rights obligations. This includes, for example, the taxation of the extraction and consumption of non-renewable resources, and forms of progressive taxation that are sensitive to the welfare of poor and low-income people (e.g., by taxing consumption of luxuries). Fiscal policy space can be further broadened by the elimination of corporate tax incentives (including tax holidays in export processing zones), and the phasing out of harmful subsidies. If the priorities are properly defined, fiscal policies can become a powerful instrument to reduce social inequalities, eliminate discrimination and promote the transition to sustainable production and consumption patterns.

The necessary reforms should not be limited to the national level. The strengthening of public finance is necessary at all levels, from the development of municipal fiscal systems and sufficient financial support for local authorities, to the provision of predictable and reliable funding to the UN system at a level sufficient to enable it to fulfill its mandates. In particular, governments should reverse the trend towards voluntary, non-core and earmarked contributions and the increasing reliance on philanthropic funding. A basic prerequisite for the strengthening of national fiscal systems is the strengthening of global tax cooperation to counter harmful tax competition and various schemes of tax avoidance and evasion.

Strengthening Public Policies Instead of Investors' Rights

Corporate lobby groups have been advocating forcefully against 'overregulation', and for the continuation of exactly those trade, investment and financial rules that have destabilized the global economy and exacerbated inequalities in both the global North and the global South. Furthermore, a new generation of free trade and investment agreements risks a further reduction in the policy space of governments to implement sound social, environmental and developmental policies. These agreements will add to the power of investors and big corporations and, by the same token, weaken the role of the State and its ability to promote human rights and sustainability. Governments should fundamentally rethink their approach towards trade and investment liberalization and take into account the demands of civil society organizations, trade unions, indigenous peoples, human rights experts and many others, to place human rights and the principles of sustainable development at the core of all trade and investment agreements. This includes the ability to implement active industrial policies to enable the rise of a strong domestic enterprise sector in countries of the global South.

Rethinking PPPs: Searching for Alternatives

Business actors and corporate think tanks like the WEF have been steadily promoting PPPs as the primary model to fill the global funding gap in infrastructure investment. Many governments have followed their advice. But as mentioned above, many studies, including those by mainstream think tanks, prove that PPPs can involve enormous risks and costs to the public sector, exacerbate inequalities and decrease equitable access to essential services. Governments should take these findings and concerns into account, rethink their approach towards private sector participation in infrastructure investment, and explore alternative means of public infrastructure financing. This may include revenues from property taxes, service charges and user fees, in compliance with human rights standards, funding by public banks, the issuance of public (including municipal) bonds, ways to cross-subsidize different public services, and, in certain cases, ODA funding.

Creating Binding Rules on Business and Human Rights and UN-Business Interactions

Experience shows that corporate social responsibility (CSR) initiatives and voluntary guidelines, such as the UN Guiding Principles on Business and Human Rights (UNGP) have failed to hold corporations accountable. Various governments, CSOs and human rights experts have concluded that there is a need for a legally binding instrument (or 'treaty') to regulate, in international human rights law, the activities of transnational corporations and other business enterprises. The Human Rights Council took a milestone decision by establishing an intergovernmental working group to elaborate such an instrument. Governments and CSOs should take this 'treaty process' seriously and engage actively in it. This process offers the historic opportunity for governments to demonstrate that they put human rights over the interests of big business. This will be a critical prerequisite for implementing the 2030 Agenda, not least the goal to ensure sustainable consumption and production patterns.

Similarly, the UN should develop a regulatory framework for UN-business interactions (including the various forms of partnerships). It should set minimum standards for the participation of the UN in global partnerships and for the shape and composition of UN initiatives involving the private sector. These standards should prevent undue corporate influence on UN policies and prevent companies that violate internationally agreed environmental, social and human rights standards or otherwise violate UN principles (via corruption, breaking UN sanctions, lobbying against UN global agreements, evading taxes, etc.) from participation in UN events and from eligibility for UN procurement.

One essential element of such a framework should be a mandatory conflict of interest and public disclosure policy for all interactions with non-State actors, with additional requirements specific to the respective UN funds, programmes and specialized agencies. Furthermore, such a regulatory framework should distinguish clearly between corporate actors and CSOs and refrain from treating fundamentally different actors as equals.

Dismantle Corporate Power and 'Too Big to Fail' Entities

The deregulation and privatization policies of the last decades have enabled increasing market concentration and the accumulation of wealth and economic power in the hands of a relatively small number of corporations and ultra-rich individuals. Existing competition and anti-trust laws have been obviously too weak to prevent mega-mergers, as recently have taken place in the agribusiness sector, and to curtail the massive growth of financial conglomerates with disproportionate influence on the global economy—and thereby directly or indirectly on the implementation of the SDGs.

In order to strengthen the role of the State and democratic decision-making processes on issues of common interest in societies, as well as ensure the provision of public services governments have to take effective measures to dismantle corporate power and prevent the further existence of corporate 'too big to fail' entities, particularly in the global shadow banking system. They should strengthen national and regional anti-trust laws, cartel offices and competition regulators. And they have to improve anti-trust policies, cooperation and legal frameworks at the global level under the auspices of the UN. This could include the development of a UN Convention on Competition, as proposed by the ETC Group.

Changing the Mindset: Reclaiming the Public Space

The measures listed above are indispensable to counteract the growing, non-monitored influence of corporate interests in the implementation of the 2030 Agenda and beyond. But these measures are not ends in themselves. There is a need to reconsider the current mainstream approach based on voluntary governance and partnerships among diverse 'stakeholders'. It is important to re-establish a clear distinction between those who should regulate and the party to be regulated and to reject any discourse that obfuscates the fact that corporations have a fundamentally different primary interest from that of governments, UN agencies, CSOs, and social movements: corporations' primary interest—enshrined in their fiduciary duty—is to satisfy the interests of their owners, creditors and shareholders. The stakeholder discourse blurs this important distinction between the different actors.

Certainly, meaningful engagement with all sectors of society is a pre-requisite for democratic decision-making as well as providing invaluable and essential expertise in the identification of problems and solutions. Governments and the UN should continue to develop their commitments and capacities in this area without relying on a one-size-fits-all approach. They should develop models which will allow all actors in society to make contributions and to protect against the influence of vested interests. Rather than continuing to 'innovate' through 'outsourcing' tasks to piecemeal partnerships with undemocratic decision-making structures, it is time for civil society to reclaim the public space—and for governments to put in place the necessary regulatory and global governance framework.

In the preamble to the 2030 Agenda governments described the "enormous disparities of opportunity, wealth and power" as one of the immense challenges (i.e., obstacles) to sustainable development.[29] The SDGs can only be achieved when governments take active political steps to overcome these disparities.

References

Alexander N (2016) Infrastructure investment and Public Private Partnerships. Washington, D.C., Heinrich Böll Stiftung North America (G20 Themes #5). https://us.boell.org/sites/default/files/uploads/2016/12/https___www.boell.de_sites_default_files_uploads_2016_12_g20-themes5-infrastructure-investment-ppp_0.pdf. Accessed 15 Jan 2018

Alstadsæter A, Johannesen N, Zucman G (2017) Tax evasion and inequality. Norwegian University of Life Sciences/University of Copenhagen/UC Berkeley and NBER. http://gabriel-zucman.eu/files/AJZ2017.pdf. Accessed 15 Jan 2018

Boys D (2017) The new generation of PPPs in infrastructure – meeting the needs of institutional investors. In: Civil Society Reflection Group on the 2030 Agenda for Sustainable Development (2017), pp 93–94

Civil Society Reflection Group on the 2030 Agenda for Sustainable Development (2017) Spotlight on Sustainable Development 2017. Reclaiming policies for the public. Privatization, partnerships, corporate capture and their impact on sustainability and inequality - assessments and alternatives. Beirut/Bonn/Ferney-Voltaire/Montevideo/New York/Penang/Rome/Suva. www.2030spotlight.org. Accessed 15 Jan 2018

Crivelli E, de Mooij RA, Keen M (2015) Base erosion, profit shifting and developing countries. IMF, Washington, D.C. www.sbs.ox.ac.uk/sites/default/files/Business_Taxation/Docs/Publications/Working_Papers/Series_15/WP1509.pdf. Accessed 15 Jan 2018

ETC Group (2017) Agribusiness mega-mergers expose need for UN Competition Convention. In: Civil Society Reflection Group on the 2030 Agenda for Sustainable Development (2017), pp 43–45

[29] UN (2015b), para. 14.

Florquin N (2011) A booming business: private security and small arms. In: Small Arms Survey (2011), pp 101–133

High-Level Panel of Eminent Persons on the post-2015 Development Agenda (2013) The Report of the High-level Panel of Eminent Persons on the Post-2015 Development Agenda. New York. http://report.post2015hlp.org/digital-report-chapter-4.html. Accessed 15 Jan 2018

Jomo KS, Chowdhury A, Sharma K, Platz D (2016) Public-Private Partnerships and the 2030 Agenda for Sustainable Development: fit for purpose? New York, UN Department of Economic & Social Affairs (DESA Working Paper No. 148, ST/ESA/2016/DWP/148). https://sustainabledevelopment.un.org/content/documents/2288desaworkingpaper148.pdf. Accessed 15 Jan 2018

Kishimoto S (2016) Remunicipalization: a practical guide for communities and policy makers. Transnational Institute, Amsterdam. www.blueplanetproject.net/wordpress/wp-content/uploads/PublicWaterForAll-Mar14-Remun.pdf. Accessed 15 Jan 2018

Oxfam America (2017) Rigged reform: US companies are dodging billions in taxes but proposed reforms will make things worse. Media briefing April 2017. www.oxfamamerica.org/static/media/files/Rigged_Reform_FINAL.pdf. Accessed 15 Jan 2018

The Economist Intelligence Unit (2015) The road from principles to practice. Today's challenges for business in respecting human rights. Geneva/London/Frankfurt/Paris/Dubai. www.economistinsights.com/sites/default/files/EIU-URG%20-%20Challenges%20for%20business%20in%20respecting%20human%20rights%20WEB_corrected%20logos%20and%20UNWG%20thx.pdf. Accessed 15 Jan 2018

UN (2015a) Addis Ababa Action Agenda of the Third International Conference on Financing for Development. New York. (UN Dok. A/RES/69/313). www.un.org/esa/ffd/wp-content/uploads/2015/08/AAAA_Outcome.pdf. Accessed 15 Jan 2018

UN (2015b) Transforming our world: the 2030 Agenda for Sustainable Development. New York (UN Doc. A/RES/70/1). https://sustainabledevelopment.un.org/post2015/transformingourworld. Accessed 15 Jan 2018

Vitali S, Glattfelder JB, Battiston S (2011) The network of global corporate control. PLoS ONE 6:10. http://arxiv.org/PS_cache/arxiv/pdf/1107/1107.5728v2.pdf. Accessed 15 Jan 2018

World Bank et al (2015) From billions to trillions – transforming development finance. Washington, D.C. http://siteresources.worldbank.org/DEVCOMMINT/Documentation/23659446/DC2015-0002(E)FinancingforDevelopment.pdf. Accessed 15 Jan 2018

World Economic Forum (2010) Everybody's business: strengthening international cooperation in a more interdependent world - report of the global redesign initiative. Geneva. http://www3.weforum.org/docs/WEF_GRI_EverybodysBusiness_Report_2010.pdf. Accessed 15 Jan 2018

Wulf A (2017) Cashing in on SDG 4. In: Civil Society Reflection Group on the 2030 Agenda for Sustainable Development (2017), pp 57–63

Jens Martens is Executive Director of Global Policy Forum (New York/Bonn) and has been the director of Global Policy Forum Europe since its foundation in 2004. Since 2011 he has coordinated the international Civil Society Reflection Group on the 2030 Agenda for Sustainable Development. From 2003 to 2009 he was member (2006–2009 Co-Chair) of the Coordinating Committee of Social Watch, a global network of several hundred NGOs working on poverty eradication and social justice. He is also a member of the Advisory Board of the Development and Peace Foundation. Previously, he was a longstanding member of the Executive Board of the German NGO World Economy, Ecology and Development (WEED). Prior to joining the staff at WEED he worked as freelance author and advisor for several NGOs and Foundations, among others the German NGO Forum on Environment and Development, the Friedrich Ebert Foundation, and the Development and Peace Foundation. From 1991 to 1992 he worked as librarian and research fellow at the German Commission for UNESCO in Bonn. He has published many articles in journals, newspapers and handbooks, has written several books and studies on development policy and UN reform, and has been co-editor of three books on German UN policy, privatization and corporate accountability. Between 2003 and 2015 he was author of the German Reality of Aid Report published annually by Deutsche Welthungerhilfe and terre des hommes Germany. He studied economics and political sciences at the University Erlangen-Nürnberg and at the Free University of Berlin.

Towards a Division of Labour for Sustainable Development: Extraterritorial Human Rights Obligations

Wouter Vandenhole

Contents

Abstract The Sustainable Development Goals (SDGs) adopted in 2015 are the new and ambitious global development agenda for the 2015–2030 period. That global agenda can only be realized through a global effort. What does that mean for the division of labour? Traditionally, fostering development has been seen as the primary responsibility of the territorial State. This chapter reflects from a human rights perspective on the role to be played by external governmental and intergovernmental actors in bringing about sustainable development. It ponders strengths and weaknesses of the right to development and extraterritorial human rights obligations, and identifies five challenges for human rights law: the legal status of the obligations to cooperate internationally; the distributive allocation of extraterritorial obligations; the triggers of extraterritorial human rights obligations; the scope of the extraterritorial obligation to cooperate for development; and the ability of human rights law to engage with strong definitions of development, which take growth agnosticism as their starting point.

W. Vandenhole (✉)
Law and Development Research Group, University of Antwerp, Antwerp, Belgium
e-mail: wouter.vandenhole@uantwerpen.be

M. Kaltenborn et al. (eds.), *Sustainable Development Goals and Human Rights*, Interdisciplinary Studies in Human Rights 5,
https://doi.org/10.1007/978-3-030-30469-0_13

1 Introduction

The Sustainable Development Goals (SDGs) adopted in 2015 are the new and ambitious global development agenda for the 2015–2030 period. That global agenda can only be realized through a global effort. What does that mean for the division of labour? Traditionally, fostering development has been seen as the primary responsibility of the territorial State, with a secondary and often more charity-based role for external actors such as the United Nations organisations, donor countries and non-governmental organisations.

This chapter reflects from a human rights perspective on the role to be played by external governmental and intergovernmental actors in bringing about sustainable development in the context of the SDGs.[1] Are States[2] that are in a position to do so, also legally obliged to provide development assistance (ODA), and to what extent: should they simply seek to meet the 0.7% GNP target, or provide development assistance commensurate to their ability or to global need? Beyond development assistance, do States have more fundamental and encompassing obligations to further sustainable development globally through structural economic and political reforms? When and how are these obligations of external actors triggered: whenever the need arises? Or only when a territorial State is unable and/or unwilling to realize development?

Answers to these questions will be sought primarily in the realm of human rights law. Over the past decades, two main human rights framings have been proposed to analyse the human rights obligations for development of external States: the right to development, and extraterritorial human rights obligations. Both framings are only emerging and major sites of political and legal contestation. Therefore, the legal materials used in this chapter show varying degrees of "legal hardness".

Section 2 explores how the 2030 Agenda for Sustainable Development and the SDGs themselves envisage the division of labour between the territorial State and external State actors. Section 3 introduces the human rights framings of obligations incumbent on external States, that is the right to development and extraterritorial obligations. Section 4 maps the areas of convergence between the Agenda for Sustainable Development and human rights law in this regard, and examines the potential contribution the latter may make in clarifying the role of external State actors in development. Section 5 highlights some of the outstanding debates in human rights law that currently prevent it from taking full normative leadership in the debate on the division of labour for sustainable development. Section 6 concludes.

[1]For reasons of space, this chapter does not engage with the question of obligations and responsibility incumbent on non-state actors, such as businesses.

[2]"States" refers to States acting individually and/or collectively through intergovernmental organisations.

2 The 2030 Agenda for Sustainable Development

Transforming our World: the 2030 Agenda for Sustainable Development, the outcome document of the United Nations summit for the adoption of the post-2015 development agenda, acknowledges and emphasizes the importance of a "Global Partnership for Sustainable Development".[3] The idea of a global partnership goes back to the Millennium Declaration and the Millennium Development Goals (MDGs), the 2000–2015 development agenda, in which MDG 8 was dedicated to the development of a global partnership for development.[4] Among the five "Ps" that the Preamble to the 2030 Agenda for Sustainable Development lists, one is referring to "Partnership". Under that heading, reference is made to "a spirit of strengthened global solidarity" as the foundation for such a partnership. In the Declaration itself, a separate heading is dedicated to the means of implementation, which a revitalized Global Partnership is considered central to.[5]

The starting point for any division of labour for sustainable development nonetheless remains that "each country has primary responsibility for its own economic and social development". Hence, external State actors have only secondary obligations for development. Another element that sheds some light on the envisaged division of labour is the principle of common but differentiated responsibilities, first established in the Rio Declaration on Environment and Development.[6] This principle of environmental law acknowledges explicitly that not all States are equal, hence the differentiated responsibilities.

Strong emphasis is put in the Declaration on financial resources, with explicit reference to the outcome document of the third International Conference on Financing for Development (held in 2015 in Addis Ababa) and to the long-standing commitment of developed countries to spend 0.7% of gross national income on official development assistance (ODA).[7] Elsewhere in the Declaration, some attention is nonetheless also paid to structural challenges in the economic system, where the need for "an enabling international economic environment, including coherent and mutually supporting world trade, monetary and financial systems, and strengthened and enhanced global economic governance" is acknowledged.[8] In line with this, the targets listed under SDG 17 relate not only to finance, but also to technology transfer, trade and systemic issues of policy and institutional coherence and multistakeholder partnerships.

In sum, primary responsibility of the territorial State is complemented with international solidarity, which is put into practice through official development assistance and an enabling international economic environment. The latter reference

---

[3]UNGA (2015).

[4]UNGA (2010) and United Nations Secretary-General (2001).

[5]UNGA (2015) at paras. 39–46, and 62.

[6]UNGA (2015), at para 12.

[7]UNGA (2015), at paras. 40 and 43.

[8]UNGA (2015), at para 63.

to macro-economic factors, read in conjunction with the principle of common but differentiated responsibilities, may suggest that structural imbalances of power, and hence different degrees of responsibility for mal- and underdevelopment, are being acknowledged.

3 International Human Rights Law

In human rights law, two main framings of obligations incumbent on external State actors have taken shape over the past decades, that is the right to development and extraterritorial obligations. Both are briefly introduced in what follows.

The Right to Development

The concept of a (human) right to development was coined and legally framed in the late 1960s and early 1970s. A Declaration on the Right to Development was adopted in 1986.[9] Unanimous political recognition to the right to development *as a human right* was given in 1993, in the Vienna Declaration and Plan of Action.[10]

Justification for the right to development was sought in the strategic, economic and political domination of the North over the global South. This domination made the North responsible for development in the South.[11]

In origin, the right to development was a radical human rights framing of responsibility for development, since it sought to challenge prevailing international economic relations, and to introduce alternative legal principles to guide international relations, such as international solidarity, substantive equality and international justice.[12] These alternative legal principles relate to the external dimension of the right to development, i.e. the claims to international cooperation and assistance a developing State may have towards external States or the international community,[13] or the right to have removed structural obstacles to development that are inherent to the prevailing international economic relations.[14] The latter aspect shows that there was a close connection between the right to development and NIEO, at least in the early years.[15]

---

[9]GA resolution, UN Doc A/RES/41/128 (4 December 1986).

[10]Art. 10 *Vienna Declaration and Programme of Action*, UN Doc A/CONF.157/23 (12 July 1993).

[11]M'baye (1972), pp. 514–526; M'baye (1980), pp. 78–88.

[12]M'baye (1980), pp. 85–88; Bedjaoui (1987), pp. 95–101; De Vey Mestdagh (1981), pp. 43–44 and 46; Alston (1980), pp. 104–105.

[13]De Feyter (1992), p. 556; Salomon (2008), pp. 2–8.

[14]Marks (1985), p. 509. See Salomon (2008), pp. 8–9.

[15]See, e.g. Abi-Saab (1980), pp. 164–168; Israel (1983), p. 29.

The 1986 Declaration tones down the radical edge of the right to development. Article 3(3) stipulates that "States have the duty to co-operate with each other in ensuring development and eliminating obstacles to development ...", but does not clarify the objectives of that cooperation. Article 4 adds that States "have the duty to take steps, individually and collectively, to formulate international development policies with a view to facilitating the full realization of the right to development". The latter Article also spells out to some extent the division of labour in bringing about development: "As a complement to the efforts of developing countries, effective international co-operation is essential in providing these countries with appropriate means and facilities to foster their comprehensive development."

Recent attempts to elaborate on the right to development's meaning in reference to the global partnership for development in MDG 8, do not add much analytical clarity.[16] The High Level Task Force set up to assist the UN open-ended working group on the right to development between 2004 and 2010,[17] elaborated an assessment framework for the extent to which States acting collectively and individually contribute to the creation of an enabling environment for the realization of the right to development.[18] It has not sought, though, to clarify the distributive allocation of obligations to the territorial State and external States, individually or collectively.[19] Even more so, in recent attempts at further standard-setting, the external dimension of the right to development, that is the obligations it imposes on external States acting individually and collectively, has been more and more downplayed.[20]

In sum, the 1986 Declaration on the Right to Development focuses mainly on international cooperation, not so much on addressing structural impediments to development. It considers the territorial State as the primary duty-bearer, but a complementary role is envisaged for external States through international cooperation, in particular through international development policies. Later attempts to operationalize the right to development under the MDGs have introduced the notion of an enabling international economic environment, but they do not shed any light on the distributive allocation of development obligations. Let us now turn to human rights law as codified in treaties, to see how responsibility for development is attributed, if at all.

[16]Vandenbogaerde (2013), pp. 187–209.

[17]The open-ended working group was established by the UN Commission on Human Rights in 1998.

[18]Report of the high-level task force on the implementation of the right to development on its sixth session (Geneva, 14–22 January 2010), Addendum—Right to development criteria and operational sub-criteria, UN Doc. A/HRC/15/WG.2/TF/2/Add.2 of 8 March 2010.

[19]For an explicit rejection of such an approach, see Report of the high-level task force on the implementation of the right to development on its sixth session (Geneva, 14–22 January 2010), Addendum—Right to development criteria and operational sub-criteria, UN Doc. A/HRC/15/WG.2/TF/2/Add.2 of 8 March 2010, para. 18.

[20]Vandenbogaerde (2013), pp. 199–200.

Extraterritorial Obligations

Typically, human rights obligations are incumbent on the territorial State, i.e. the State that exercises territorial jurisdiction. Nonetheless, there is a longstanding recognition that the realisation of at least some human rights—in particular economic, social and cultural rights (ESC rights), children's rights, and the rights of persons with disabilities—requires international cooperation and assistance.[21] The precise meaning and scope of these references to international cooperation and assistance remain somewhat open to debate,[22] although a strong case has been made for reading extraterritorial obligations into these references.

The position of the UN Committee on Economic, Social and Cultural Rights (CESCR) on extraterritorial obligations under the ICESCR has been consistent over the past two decades,[23] and is well-reflected, for example, in its 2016 general comment on the right to sexual and reproductive health. Territorial States that are not able to comply with their obligations due to a lack of resources must seek international cooperation and assistance. "States that are in a position to do so must respond to such requests in good faith and in accordance with the international commitment of contributing at a minimum 0.7 per cent of their gross national income for international cooperation and assistance." Donor States and international actors have an obligation to comply with human rights standards throughout their international assistance (in other words, it should be rights-based), and as members of international organisations and in bilateral, regional and multilateral treaties.[24] Emphasis was put in an earlier general comment on the right to health on the role of international assistance and cooperation in realizing core obligations and obligations of comparable priority.[25]

Overall, the interpretation of the Committee on the Rights of the Child (CRC Committee) has been very much in line with that of the CESCR,[26] as illustrated for example in the former's general comment on public budgeting. States that lack the

---

[21]See Article 2 International Covenant on Economic, Social and Cultural Rights; Article 4 Convention on the Rights of the Child; Articles 4 and 32 Convention on the Rights of Persons with Disabilities.

[22]An overview of the UN human rights bodies' understanding can be found in a working paper published by the Global Initiative for Economic, Social and Cultural Rights, Human Rights Law Sources: UN Pronouncements on Extra-Territorial Obligations, July 2015, see http://www.etoconsortium.org/nc/en/main-navigation/library/documents/?tx_drblob_pi1%5BdownloadUid%5D=163 (last accessed 5 January 2018).

[23]For a systematic overview, see Coomans (2011), pp. 1–35.

[24]CESCR, GC No. 22 (2016) on the right to sexual and reproductive health (article 12 of the International Covenant on Economic, Social and Cultural Rights), UN Doc E/C.12/GC/22, 2 May 2016, paras. 50–52.

[25]Social and Cultural Rights Committee on Economic, 'General Comment No. 14 (2000) on the Right to the Highest Attainable Standard of Health (Article 12 of the International Covenant on Economic, Social and Cultural Rights)', (11 August 2000) at para. 45.

[26]For a comprehensive and systematic analysis, see Vandenhole (2009), pp. 23–63.

resources to fully realize the CRC must seek international cooperation; States with resources "have an obligation to provide such cooperation with the aim of facilitating the implementation of children's rights in the recipient State."[27] Beyond bilateral development assistance, States "should collaborate with other States' efforts to mobilize the maximum available resources for children's rights."[28] Development cooperation strategies should be rights-based: they should not negatively impact on children, and should contribute to the realisation of children's rights.[29] State must observe their children's obligations "when engaging in development cooperation as members of international organizations, [footnote omitted] and when signing international agreements."[30]

The Committee on the Rights of Persons with Disabilities (CRPD Committee) has mainly emphasized that all cooperation must be rights-based. With regard to the right to inclusive education, this means that "all bilateral and multilateral cooperation must aim to advance inclusive and equitable quality education and promote lifelong learning opportunities for all [. . .]."[31]

In sum, the importance of international assistance and cooperation for the realisation of human rights has been acknowledged in several human rights treaties. The reference to international assistance and cooperation has been understood by the UN human rights treaty bodies as imposing an obligation for developing countries to seek, and for developed countries to offer development assistance. Development assistance, international agreements and the exercise of membership of international organizations should all be rights-based. Again, there is little if any clarification of the division of labour for sustainable development. In the next two sections, I will spell out in more detail some aspects of the extraterritorial human rights obligations framing, while looking into in the convergence of the SDGs and human rights, and the challenges ahead.

4 Convergence and Added Value of Human Rights

On a number of key points, the 2015 Agenda and the SDGs on the one hand, and human rights law on the other hand, converge and mutually reinforce each other.

First, the need for international cooperation, including international assistance, is acknowledged: development is not exclusively the responsibility of the territorial State. The 2015 Agenda and the SDGs mainly refer to solidarity and a global

[27]Committee on the Rights of The Child (2016) at para. 35.

[28]Committee on the Rights of The Child (2016), at para. 37.

[29]Committee on the Rights of The Child (2016), at para. 38.

[30]Committee on the Rights of The Child (2016), at para. 39.

[31]CRPD, General comment No. 4 (2016) on the right to inclusive education, UN Doc CRPD/C/GC/4, 25 November 2016, para. 43. Compare CRPD, GC comment No. 2 (2014) Article 9: Accessibility, UN Doc CRPD/C/2, 22 May 2014, para. 47.

partnership, and seem to envisage mainly multilateral cooperation. The same holds true to some extent for the right to development. The extraterritorial human rights obligations frame focuses more on bilateral cooperation, and in particular on development assistance, although it also pays attention to States acting collectively through international organisations or a treaty regime.

Second, the primary responsibility for development (understood respectively as sustainable development or the full realisation of human rights) lies with the territorial State. External State actors have only a secondary role to play. There is also some differentiation of responsibility: the 2015 Agenda speaks about "shared but differentiated responsibilities" in language borrowed from the Rio Declaration; in the human rights context, the obligations of developing and developed states are different: the latter have to step in when the former are unable to fully realize their obligations and ask the latter for assistance.

Third, it is acknowledged, albeit in various degrees, that development is not just a matter of development finance or development assistance. Attention is given also to economic and financial governance, and to the need for an enabling international economic environment.

Beyond these commonalities, human rights law may be able to further refine the emerging picture of the division of labour for sustainable development in three respects. First, it locates international cooperation more explicitly in the realm of obligation than of charity. Second, it contains more detail on the respective obligations of developing and developed States. And finally, the tripartite typology of human rights obligations may help to clarify the obligations incumbent on external actors in realizing sustainable development. That it *may* be able is not to say that it currently *is* already fully up to that task.

Human rights law locates international cooperation and assistance stronger in the realm of (legal) obligation rather than in that of charity. In Fukuda-Parr's view, the challenge for international development policy (at the time as reflected in MDG 8) as already done in human rights law was to "shift international cooperation from charity to solidarity".[32] Whether that shift has already fully taken place in human rights law, is questionable though. During negotiations on new human rights instruments, this has been a recurrent topic of contestation and politicization.[33] We will return to this point in Sect. 5.

Second, human rights law spells out in somewhat more detail than the 2015 Agenda and the SDGs what the respective obligations of the territorial and external States are. The territorial State has to undertake measures to the maximum of its available resources to (progressively) fully realize the economic, social and cultural rights guaranteed by the treaties concerned. Where needed, it is under an obligation

[32]Fukuda-Parr (2006), p. 967.

[33]With regard to the negotiations on the Optional Protocol to the ICESCR, see Vandenbogaerde and Vandenhole (2010), pp. 227–230. With regard to the negotiations on the CRPD, see Vandenhole (2009), pp. 55–60.

to seek international assistance and cooperation.[34] In a spirit of fair burden-sharing, the strongest shoulders must bear the heaviest burdens: "economically developed States parties have a special responsibility".[35] Hence, it is "particularly incumbent upon those States which are in a position to assist"[36] to respond to any request for assistance. In Sect. 5, we will explore what the precise legal status of this obligation is.

Third, the tripartite typology of obligations (to respect, to protect, and to fulfil), when applied to extraterritorial obligations, takes the debate beyond official development assistance and the transfer of financial resources, to more structural factors that inhibit development or create underdevelopment or mal-development. This may be helpful to deepen the understanding of what it takes to create an enabling international economic environment. As the CESCR nicely explains in its 2017 general comment on State obligations in the context of business activities, the extraterritorial obligation to respect requires States "not to obstruct another State from complying with its obligations [...]. This duty is particularly relevant to the negotiation and conclusion of trade and investment agreements or of financial and tax treaties [...]",[37] and—one could add—when exercising membership in regional or international organizations. The extraterritorial obligation to protect implies that the home state of companies has to protect individuals in other countries against adverse impacts of these companies, through e.g. regulation and the use of incentives.[38] The extraterritorial obligation to fulfil may include an obligation to live up to the longstanding commitment to spend 0.7% of GNI on official development assistance, but goes clearly beyond that. For CESCR, with regard to State obligations in the context of business activities, it seems to coincide with promoting and creating an international enabling environment.[39]

The account offered here of the reinforcing and clarifying role that human rights law can play with regard to the division of labour for sustainable development may be too optimistic. Much will depend on the way and extent to which human rights law evolves in the coming years. I now turn to the main challenges ahead.

[34]Article 2 International Covenant on Economic, Social and Cultural Rights; Article 4 Convention on the Rights of the Child; Article 4(2) Convention on the Rights of Persons with Disabilities.

[35]Committee on Economic, Social and Cultural Rights (2000), para. 40.

[36]Committee on Economic, Social and Cultural Rights (1990) at para. 14.

[37]Committee on Economic, Social and Cultural Rights (2017) at para. 34.

[38]Committee on Economic, Social and Cultural Rights (2017), at para. 36.

[39]Committee on Economic, Social and Cultural Rights (2017), at para. 40.

5 Challenges Ahead for Human Rights Law

There are at least five challenges or uncertainties to be addressed in order to make human rights law indeed instrumental in carving out a clear division of labour for sustainable development between territorial States and external actors. These are: the legal status of the obligations to cooperate internationally; the distributive allocation of extraterritorial obligations; the triggers of extraterritorial human rights obligations; the scope of the extraterritorial obligation to cooperate for development; and the ability of human rights law to engage with strong definitions of development, which take growth agnosticism as their starting point.

Aspirational or Hard Legal Obligations?

There is a long-standing, very often highly politicised and polarised discussion whether or not there is a hard legal obligation to provide international assistance and cooperation, and in particular development aid. In doctrine, two schools have been identified by Karimova: a maximalist and a minimalist one. The maximalist school submits, on the basis of the references to international assistance and cooperation that can be found in treaties, that hard legal extraterritorial obligations exist (for some only negative obligations, for others also positive obligations). In the minimalist approach, international cooperation is not seen as a self-standing and autonomous obligation, but rather as a means or a measure by which obligations can be fulfilled.[40]

Whereas this doctrinal disagreement has not struck much of a chord with the UN treaty bodies, it does correspond to a political disagreement that exists very much along North-South lines. Inevitably, as long as this legal and political disagreement persists, the role of human rights law in carving out a division of labour for sustainable development will be somewhat constrained. In my own reading, informed by more recent treaty negotiations, there is currently no *general and undifferentiated* hard legal obligation to cooperate internationally for development. On the other hand, an extraterritorial obligation to *respect* human rights is fairly well accepted.[41]

---

[40] Karimova (2014), pp. 169–171.

[41] See also Vandenhole (2009).

The Distributive Allocation of Obligations Remains Unsettled

The distributive allocation of obligations plays at two levels: first of all, what the obligations are of the territorial State, and of external States. And secondly, how obligations are allocated *among* external States.

There is a clear divergence of views on the first question, i.e. the distributive allocation of human rights obligations to the territorial State and external States. For some, the starting point is that all States have global obligations. In their view, human rights are universal: "all people have human rights and [...] all States have the responsibility to protect those rights – for all people".[42] Hence, the key question is whether the human rights of individuals are respected[43] or, from a violations perspective, whether "the global order in its entirety" may be to blame for structural impediments to the realisation of human rights.[44] Whereas this is an interesting theoretical position that deserves further theorisation, it can hardly be grounded in current human rights law.

Another strand of thought considers the territorial State as the primary duty-bearer, and external States as secondary or subsidiary duty-bearers.[45] This position accords well with the fact that the territorial State holds sovereign power over its territory. However, I submit that the obligations of external States are not merely secondary or subsidiary. Rather, they are *complementary*: these obligations accompany but never replace a territorial State's obligation. The term "secondary" suggests a hierarchy or sequence of obligations, as if extraterritorial human rights obligations arise only in the event that the territorial State failed to observe its obligations. "Complementary" recognises that a State's extraterritorial obligations co-exist simultaneously with the territorial State's obligations, whereas it accepts that the latter's obligations remain primary.[46]

Taking the analysis one step further, it is suggested to differentiate the distributive allocation of obligations between the territorial State and external States with the type of obligation. Since profound resource implications, including financial, become more likely as we move on the obligations continuum from respect and protect to fulfil, I consider the extraterritorial obligations to respect and to protect as complementary and simultaneous, and the extraterritorial obligation to fulfil as subsidiary.[47] Simultaneous obligations are immediate obligations, lack of observance of which cannot be justified by invoking non-abidance with the primary

---

[42]Gibney (Reference) at 47.

[43]Skogly (2006), p. 48.

[44]See also Salomon (2006), pp. 113, 116; Pogge (2002).

[45]See, for example, Craven (1995), p. 144; Coomans (2004), p. 198; Sepúlveda (2003), pp. 370–377 (implicitly).

[46]Compare Vandenhole and Benedek (2013), pp. 332–366.

[47]Compare Alston's notion of a correlative obligation in case a developing country has demonstrated its best efforts to meet the Millennium Development Goals (MDGs) but is unable to do so because of a lack of financial resources, Alston (2005), pp. 755–829, at 778.

obligation by the territorial State. The extraterritorial obligation to fulfil (like the territorial State's obligation to fulfil) fundamentally requires the mobilisation of resources, and is therefore in principle not an immediate obligation, but one that is subject to progressive realisation in light of the (maximum use of) available resources.[48]

If, in a specific case, the resource implications assumption with regard to the extraterritorial obligations to fulfil, or the relatively resource-free implications assumption underlying the extraterritorial obligations to respect and to protect, can be rebutted, the assessment of the subsidiary or simultaneous nature of the extraterritorial obligations may have to be adjusted.

The division of responsibility *between* external States, in particular in the field of international development co-operation, is very much a *de lege ferenda* exercise. No particular problems arise with regard to the obligation to respect, which is a negative obligation that is simultaneously incumbent on (each territorial State and) all external States. The obligation to protect is also of simultaneous application, but may have more far-reaching implications for some States commensurate to their capacity. The obligation to protect may sometimes amount to a *joint* obligation, that is when individual extraterritorial obligations would not be sufficient.[49] This may be the case when policies of international financial institutions or of the European Union are considered to be an unjustifiable interference with ESC rights. With regard to the division of responsibility amongst external States under the extraterritorial obligation to fulfil, it is not sufficient to identify "a generic [obligation] that attaches to the undifferentiated international community".[50] Duty-bearers are to be identified in more detail. Whether a State is in a position to assist depends *inter alia* on the availability of resources, be they human, natural, financial or other.[51] This is not to say that the extraterritorial obligation to fulfil never requires collective or joint observance in order to be effective. However, as the obligation to fulfil is most likely to have some resource implications, even when considered to be a joint obligation of a number of States, it is necessary to specify which external States are concerned and what the extent of their obligations is. In order to be able to do so, further clarification is needed on how extraterritorial obligations of States are triggered.

---

[48]Certain exceptions to the principle of progressive realisation may apply, such as non-discrimination and core obligations. For more details, see, for example, Vandenhole (2003), pp. 437–438 and 441–442.

[49]Künnemann (2004), p. 220.

[50]See Alston (2005), p. 777: "The nature of any obligation that could reasonably be argued to have emerged from the various commitments to cooperate internationally is, at best, a generic one that attaches to the undifferentiated international community."

[51]Engh (2008).

Triggers of Extraterritorial Fulfil Obligations Need to Be Defined More Clearly

A diversity of triggers can be considered. It may be the mere ability of the external State(s), as proponents of global human rights obligations suggest. Or it may be the request by a territorial State that is unable to realize human rights on its own. Other triggers that have been suggested are causation of human rights harm,[52] or historical responsibility.

The domestic State's inability, for reasons beyond its control, to abide by its obligation to fulfil the minimum core content of ESC rights would correspond partly to the CESCR's threshold for activating the obligation to fulfil (provide) for a territorial State.[53] The extraterritorial obligation to fulfil, like the domestic obligation to fulfil, is subject to progressive realisation in light of the maximum of available resources. It is, in other words, not an immediate obligation, but rather one qualified by time and resources. However, upon the receipt of a legitimate request, there is an immediate obligation for States in a position to assist to take steps to contribute to the progressive realisation in the affected State.

The *applicability* of extraterritorial obligations of fulfilment should not be conditioned by the reasons for noncompliance by the territorial State. Unwillingness or lack of commitment does not necessarily exonerate other States from taking action.[54] This does not make external States the *de facto* primary duty-bearers. The territorial State remains responsible for noncompliance with its obligation to fulfil ESC rights to the maximum extent of its available resources. The action taken by external States reflects their living up to their part of responsibility: their extraterritorial obligations of fulfilment remain subsidiary, constrained by the availability of resources and arguably limited to the core obligations.[55] Hence, the *applicability* of extraterritorial obligations of fulfilment should not be conditioned by whether the territorial State fully complies with its own primary obligations, regardless of the reasons for this noncompliance (be it inability or unwillingness). However, their *scope* is not

---

[52]Skogly (2013), pp. 233–258.

[53]See, for example, Committee on Economic, Social and Cultural Rights (1999), para 15: "whenever an individual or group is unable, for reasons beyond their control, to enjoy the right to adequate food by the means at their disposal, States have the obligation to fulfil (provide) that right directly". The reference to "beyond their control" was borrowed from Art. 25(1) of the Universal Declaration on Human Rights.

[54]Compare on the external obligations of a supranational organization like the European Union (EU), Hazelzet (2007), p. 401: "It is argued, moreover, that the EU's external duties, for instance in the area of trade and development, should not be conditional on the fulfilment of obligations by states in the South, because the EU's international assistance and cooperation, ultimately, are to the people in the South, with the state being the intermediary through which their rights can most often be realised."

[55]As the CESCR has repeatedly suggested, see Committee on Economic, Social and Cultural Rights (2002) at para. 38; Committee on Economic, Social and Cultural Rights (2006) at para. 40; Committee on Economic, Social and Cultural Rights (2008), para. 61.

extended when a territorial State fails to live up to its part. For example, the fact that a territorial State is reluctant to provide free basic health care to all individuals does not exonerate other States from fulfilling their extraterritorial obligations in this regard, albeit limited to their share. How the scope of the extraterritorial obligation to fulfil-provide can be defined, is discussed in the next sub-section.

Scope of the Obligation to Cooperate for Development

The scope of the extraterritorial obligation to fulfil-provide can be defined by the resources available to a State. Resources include not only financial, but also natural and human resources, as well as, for example, technology, logistics and information.[56] Nonetheless, financial resources will often play an important role.

Availability of resources is primarily assessed by the State concerned, which enjoys a margin of appreciation in the allocation of its resources. A *prima facie* assessment of *financial* resources an external State has available to cooperate for development may be based primarily on the allocations actually made in the national budget for ODA and for emergency aid. A more comprehensive assessment should go beyond national allocations. For example, political or legal commitments made in internal development policies, domestic legislation or bilateral agreements may all be indicative of the financial resources an external State can be expected to mobilize under its extraterritorial obligation to fulfil-provide. One can think there of the oft-repeated political commitment of spending 0.7% of GNI on development assistance. Other commitments made, for example, in the context of conferences regarding financing for development can equally help to settle the scope of the extraterritorial obligation to fulfil, both as a collective obligation and as one incumbent on particular States.

In its *Statement on Poverty*, CESCR submitted that "core obligations give rise to ... international responsibilities for developed States, as well as others that are 'in a position to assist'".[57] Core obligations hence seem to have been elevated to the level of global human rights obligations.

Extraterritorial human rights obligations to co-operate internationally for development should not be drawn too narrowly, though. Development assistance is but one feature of the obligation to co-operate internationally for development. Beyond the obligation to fulfil-provide, all types of obligations may be relevant. For example, as part of the extraterritorial obligation to respect, a human rights impact assessment may have to be undertaken of all policies with external effects. As part of the obligation to fulfil-facilitate, structural measures in trade, finance and investment regimes may be needed.

[56]Compare Engh (2008).

[57]Committee on Economic, Social and Cultural Rights (2001), para. 16.

Sustainable Development and Growth Agnosticism

A fifth challenge for human rights law is to take sustainable development, particularly in its strong definitions, seriously.

In the 2030 Agenda, economic growth, social development and ecological sustainability go hand in hand. Commonly, sustainable development is explained in terms of a triple bottom-line (the three Ps of profit, planet and people), suggesting that it is about *balancing* economic growth with environmental and social considerations. In other words, economic growth is still taken for granted.

This approach of *balancing* planet and people with profit may be seen as the weak definition of sustainability. Strong definitions of sustainability deplore that sustainable development means "environmentally friendly economic growth".[58] In these strong definitions, the *prevalence* of the environmental dimension over the economic one is argued for.[59] The prioritization of the environmental (*planet*) over the economic (*profit*) one[60] has led to a radical departure from assumptions of economic growth, including zero-growth or even de-growth, as argued in post-growth or ecological economics.[61] At the very least, growth agnosticism, as adopted by Raworth in "Doughnut Economics",[62] replaces growth as the new normal.

At first sight, growth agnosticism seems to undercut the foundation of social development. If economic growth is no longer the basis of social development, what is? Or does growth agnosticism lead to a major roll-back in socio-economic rights realization? The latter is certainly not the option that is defended. In "Doughnut Economics", it is between the outer boundary, i.e. the ecological ceiling (composed of Roxtrom et al.'s nine planetary boundaries) and the inner boundary, i.e. the social foundation (twelve social boundaries drawing on "internationally agreed minimum social standards"), that "lies an environmentally safe and socially just space in which humanity can thrive."[63] The basis for that environmentally sustainable and socially just space is to be found in redistribution, domestically and globally.

For sure, the call for global redistribution and fair shares of effort reinforces the earlier analysis in this chapter that if the SDGs are to become a reality, redistribution of some sort through international assistance and cooperation will be needed. But growth agnosticism also challenges human rights law, since the latter is very much based on assumptions of growth that translate into progressive realisation of ESC

---

[58]Kerschner (2010), p. 549.

[59]Bosselmann (2011), p. 54; compare the argument that the SDGs need to be organized 'in a normative hierarchy, with a single priority sustainability goal at the apex', see Kim and Bosselmann (2015), p. 198; on the nine planetary boundaries, see the Stockholm resilience Centre http://www.stockholmresilience.org/research/planetary-boundaries.html (accessed July, 3 2017).

[60]Bosselmann (2011); Ross (2009), pp. 32–54.

[61]For arguments about the desirability and feasibility of degrowth, see the references in Kallis et al. (2012), pp. 173–175.

[62]Raworth (2017b), pp. 243–285.

[63]Raworth (2017a).

rights. This requires a major shift in attention towards redistribution and redistributive equality, concepts that so far have remained un- or under theorised in human rights law. As submitted by Alston, the human rights community has closed it eyes to what has been coined extreme or radical inequality,[64] and it remains to be seen whether human rights law can do more than offering "a floor of protection against indigence", that is whether it can provide "a ceiling on inequality", as Moyne has pleaded for.[65]

6 Conclusions

The ambition of this chapter was to show how human rights law can be of help to the SDGs to operationalise the global partnership for development. That global agenda can only be realized through a global effort. What does that mean for the division of labour? Traditionally, fostering development has been seen as the primary responsibility of the territorial State, with a secondary and often more charity-based role for external actors such as the United Nations organisations, donor countries and non-governmental organisations. That legacy is reflected in the SDG Declaration.

Whereas the concept of the right to development was meant to address structural impediments to development, the 1986 Declaration on the Right to Development focuses mainly on international cooperation. Later attempts to operationalize the right to development under the MDGs have introduced the notion of an enabling international economic environment, but they do not shed any light on the distributive allocation of development obligations.

The reference to international cooperation and assistance in human rights treaties has been interpreted by the UN human rights treaty bodies as imposing an obligation for developing countries to seek, and for developed countries to offer development assistance. Development assistance, international agreements and the exercise of membership of international organizations should all be rights-based. Here too, there is little if any clarification of the division of labour for sustainable development.

So, beyond the areas of convergence in the SDGs and human rights law, at least five challenges or uncertainties must be addressed in order to make human rights law indeed instrumental in carving out a clear division of labour for sustainable development between territorial States and external governmental actors. These are: the legal status of the obligations to cooperate internationally; the distributive allocation of extraterritorial obligations; the triggers of extraterritorial human rights obligations; the scope of the extraterritorial obligation to cooperate for development; and the ability of human rights law to engage with strong definitions of development,

[64] Alston (2017): 'Extreme inequality should also be seen as a cause for shame on the part of the international human rights movement. Just as global economic institutions have eschewed human rights, so too have the major human rights groups avoided tackling the economics of rights.'

[65] Moyne (2015, 2017).

which take growth agnosticism as their starting point. The jury is out whether human rights law will evolve in these directions.

References

Abi-Saab G (1980) The legal formulation of a right to development. In: Dupuy R-J (ed) Le droit au développement au plan international. Nijhoff, The Hague, pp 159–175

Alston P (1980) The right to development at the international level. In: Dupuy R-J (ed) Le droit au développement au plan international. Nijhoff, The Hague, pp 99–114

Alston P (2005) Ships passing in the night: the current state of the human rights and development debate seen through the lens of the Millennium Development Goals. Hum Rights Q 27 (3):755–829

Alston P (2017) Extreme inequality as the antithesis of human rights. Opendemocracy.net

Bedjaoui M (1987) Some unorthodox reflections on the "right to development". In: Snyder F, Slinn P (eds) International law of development: comparative perspectives. Professional Books, Abingdon, pp 87–116

Bosselmann K (2011) A vulnerable environment: contextualising law with sustainability. J Hum Rights Environ 2(1):45–63

Committee on Economic, Social and Cultural Rights (1990) General Comment No. 3: The nature of States parties obligations (Art. 2, par. 1)

Committee on Economic, Social and Cultural Rights (1999) General Comment 12, Right to Adequate Food (Twentieth Session, 1999), UN Doc. E/C.12/1999/5

Committee on Economic, Social and Cultural Rights (2000) General Comment No. 14 (2000) on the right to the highest attainable standard of health (article 12 of the International Covenant on Economic, Social and Cultural Rights)

Committee on Economic, Social and Cultural Rights (2001) Statement on Substantive Issues Arising in the Implementation of the International Covenant on Economic, Social and Cultural Rights: Poverty and the International Covenant on Economic, Social and Cultural Rights (4 May 2001), UN Doc. E/C.12/2001/10

Committee on Economic, Social and Cultural Rights (2002) General Comment No. 15, The right to water (arts. 11 and 12)

Committee on Economic, Social and Cultural Rights (2006) General Comment 17, The Right of Everyone to Benefit from the Protection of the Moral and Material Interests Resulting from any Scientific, Literary or Artistic Production of Which He or She Is the Author

Committee on Economic, Social and Cultural Rights (2008) General Comment 19, The Right to Social Security (Art. 9)

Committee on Economic, Social and Cultural Rights (2017) General Comment No. 24 on State Obligations under the International Covenant on Economic, Social and Cultural Rights in the Context of Business Activities

Committee on the Rights of the Child (2016) General Comment No. 19 (2016) on public budgeting for the realization of children's rights (art. 4)

Coomans F (2004) Some remarks on the extraterritorial application of the international covenant on economic, social and cultural rights. In: Coomans F, Kamminga MT (eds) Extraterritorial application of human rights treaties. Intersentia, Antwerp

Coomans F (2011) The extraterritorial scope of the international covenant on economic, social and cultural rights in the work of the United Nations Committee on economic, social and cultural rights. Hum Rights Law Rev 11(1):1–35

Craven M (1995) The international covenant on economic, social, and cultural rights: a perspective on its development. Oxford University Press, Oxford

De Feyter K (1992) The human rights approach to development. Universitaire Instelling Antwerpen

de Vey Mestdagh K (1981) The right to development. Neth Int Law Rev 28(01):30–53
Engh I-E (2008) Developing capacity to realizse socio-economic rights. The right to food in the context of HIV/AIDS in South Africa and Uganda. Intersentia, Antwerp
Fukuda-Parr S (2006) Millennium Development Goal 8: indicators for international human rights obligations? Hum Rights Q 28(4):966–997
Hazelzet H et al (2007) New human rights duty-bearers: a conversation on policy implications. In: Salomon M, Tostensen A, Vandenhole W (eds) Casting the net wider: human rights, development and new duty-bearers. Intersentia, Antwerp, pp 395–416
Israel J-J (1983) Le droit au développement. Revue Générale de Droit International Public 87:6–41
Kallis G, Kerschner C, Martinez-Alier J (2012) The economics of degrowth. Ecol Econ 84:172–180
Karimova T (2014) The nature and meaning of 'International Assistance and Cooperation' under the international covenant on economic, social and cultural rights. In: Riedel E, Giacca G, Golay C (eds) Economic, social, and cultural rights in international law. Contemporary issues and challenges. Oxford University Press, Oxford, pp 163–192
Kerschner C (2010) Economic de-growth vs. steady-state economy. J Clean Prod 18(6):544–551
Kim RE, Bosselmann K (2015) Operationalizing sustainable development: ecological integrity as a grundnorm of international law. Rev Eur Comp Int Environ Law 24(2):194–208
Künnemann R (2004) Extraterritorial application of the international covenant on economic, social and cultural rights. In: Coomans F, Kamminga MT (eds) Extraterritorial application of human rights treaties. Intersentia, Antwerp
M'Baye K (1972) Le droit au développement comme un droit de l'homme. Revue des droits de l'homme V(2–3):505–534
M'Baye K (1980) Le droit au développement. In: Dupuy R-J (ed) Le droit au développement au plan international. Nijhoff, The Hague, pp 72–93
Marks S (1985) Emerging human rights: a new generation for the 1980s? In: Falk R, Kratochwil F, Mendlovitz S (eds) International law: a contemporary perspective. Westview, Boulder, pp 501–513
Moyne S (2015) Human rights and the age of inequality. opendemocracy.net (2017)
Pogge T (2002) World poverty and human rights: cosmopolitan responsibilities and reforms
Raworth K (2017a) The Doughnut of social and planetary boundaries (2017). <https://www.kateraworth.com/doughnut/>. Accessed 3 July 2017
Raworth K (2017b) Doughnut economics. Seven ways to think like a 21st-century economist. Random House Business Books, London
Ross A (2009) Modern interpretations of sustainable development. J Law Soc 36(1):32–54
Salomon M (2006) International human rights obligations in context: structural obstacles and the demands of global justice. In: Andreassen BA, Marks SP (eds) Development as a human right: legal, political, and economic dimensions. Harvard University Press, Cambridge, MA
Salomon M (2008) Legal cosmopolitanism and the normative contribution of the right to development. LSE Law, Society and Economy Working Papers 16/2008. London School of Economics and Political Science, London
Secretary-General, United Nations (2001) Road map towards the implementation of the United Nations Millennium Declaration
Sepúlveda M (2003) The nature of the obligations under the international covenant on economic, social and cultural rights. Intersentia, Antwerp
Skogly S (2006) Beyond National borders. States' human rights obligations in international cooperation. Intersentia, Antwerp
Skogly S (2013) Causality and extraterritorial human rights obligations. In: Langford M et al (eds) Global justice, state duties: the extraterritorial scope of economic, social and cultural rights in international law. Cambridge University Press, New York, pp 233–258
UNGA (2010) United Nations Millennium Declaration
UNGA (2015) Transforming our world: the 2030 Agenda for Sustainable Development
Vandenbogaerde A (2013) The right to development in international human rights law: a call for its dissolution. Neth Q Hum Rights 31(2):187–209

Vandenbogaerde A, Vandenhole W (2010) The optional protocol to the international covenant on economic, social and cultural rights: an ex ante assessment of its effectiveness in light of the drafting process. Hum Rights Law Rev 10(2):207–237

Vandenhole W (2003) Completing the UN complaint mechanisms for human rights violations step by step: towards a complaints procedure to the international covenant on economic, social and cultural rights. Neth Q Hum Rights 21:423–462

Vandenhole W (2009) Economic, social and cultural rights in the CRC: is there a legal obligation to cooperate internationally for development? Int J Child Rights 17:23–63

Vandenhole W, Benedek W (2013) Extraterritorial human rights obligations and the north-south divide. In: Langford M et al (eds) Global justice, state duties: the extraterritorial scope of economic, social and cultural rights in international law. Cambridge University Press, New York, pp 332–366

Wouter Vandenhole teaches human rights law at the University of Antwerp. He chairs the Law and Development research group at the Faculty of Law. Mr. Vandenhole graduated in law at the University of Leuven (Belgium), received an LL.M. in Law in Development from the University of Warwick (United Kingdom) and holds a PhD of the University of Leuven. He has previously held positions at the University of Leuven, Tilburg University (the Netherlands) and the European Master's Degree in Human Rights and Democratisation (Venice, Italy). His research interests include children's rights, human rights, in particular economic, social and cultural rights, and the relationship between human rights law and development. For some years now, he has focused on transnational human rights obligations, i.e. the human rights obligations of new duty-bearers, and in particular on companies.

The manufacturer's authorised representative in the EU is Springer Nature Customer Service Centre GmbH, Europaplatz 3, 69115 Heidelberg, Germany. If you have any concerns regarding our products, please contact ProductSafety@springernature.com

Printed and bound by CPI Group (UK) Ltd, Croydon, CR0 4YY
30/09/2026
02225652-0010